Learn, Practice, Succeed

Eureka Math®
Grade 7
Module 6

Students, families, and educators:

Thank you for being part of the *Eureka Math*® community, where we celebrate the joy, wonder, and thrill of mathematics.

In *Eureka Math* classrooms, learning is activated through rich experiences and dialogue. That new knowledge is best retained when it is reinforced with intentional practice. The *Learn, Practice, Succeed* book puts in students' hands the problem sets and fluency exercises they need to express and consolidate their classroom learning and master grade-level mathematics. Once students learn and practice, they know they can succeed.

What is in the Learn, Practice, Succeed *book?*

Fluency Practice: Our printed fluency activities utilize the format we call a Sprint. Instead of rote recall, Sprints use patterns across a sequence of problems to engage students in reasoning and to reinforce number sense while building speed and accuracy. Sprints are inherently differentiated, with problems building from simple to complex. The tempo of the Sprint provides a low-stakes adrenaline boost that increases memory and automaticity.

Classwork: A carefully sequenced set of examples, exercises, and reflection questions support students' in-class experiences and dialogue. Having classwork preprinted makes efficient use of class time and provides a written record that students can refer to later.

Exit Tickets: Students show teachers what they know through their work on the daily Exit Ticket. This check for understanding provides teachers with valuable real-time evidence of the efficacy of that day's instruction, giving critical insight into where to focus next.

Homework Helpers and Problem Sets: The daily Problem Set gives students additional and varied practice and can be used as differentiated practice or homework. A set of worked examples, Homework Helpers, support students' work on the Problem Set by illustrating the modeling and reasoning the curriculum uses to build understanding of the concepts the lesson addresses.

Homework Helpers and Problem Sets from prior grades or modules can be leveraged to build foundational skills. When coupled with *Affirm*®, *Eureka Math*'s digital assessment system, these Problem Sets enable educators to give targeted practice and to assess student progress. Alignment with the mathematical models and language used across *Eureka Math* ensures that students notice the connections and relevance to their daily instruction, whether they are working on foundational skills or getting extra practice on the current topic.

Where can I learn more about Eureka Math *resources?*

The Great Minds® team is committed to supporting students, families, and educators with an evergrowing library of resources, available at eureka-math.org. The website also offers inspiring stories of success in the *Eureka Math* community. Share your insights and accomplishments with fellow users by becoming a *Eureka Math* Champion.

Best wishes for a year filled with "aha" moments!

Jill Diniz

Jill Diniz
Chief Academic Officer, Mathematics
Great Minds

Contents

Module 6: Geometry

Opening Exercise

As we begin our study of unknown angles, let us review key definitions.

Term	Definition
	Two angles, $\angle AOC$ and $\angle COB$, with a common side $\overrightarrow{OC}$, are _____ angles if C is in the interior of $\angle AOB$.
	When two lines intersect, any two non-adjacent angles formed by those lines are called _____ angles, or _____ _____ angles.
	Two lines are _____ if they intersect in one point, and any of the angles formed by the intersection of the lines is 90°. Two segments or rays are _____ if the lines containing them are _____ lines.

Complete the missing information in the table below. In the *Statement* column, use the illustration to write an equation that demonstrates the angle relationship; use all forms of angle notation in your equations.

Angle Relationship	Abbreviation	Statement	Illustration
Adjacent Angles		The measurements of adjacent angles add.	
Vertical Angles		Vertical angles have equal measures.	

EUREKA MATH®

Angles on a Line		If the vertex of a ray lies on a line but the ray is not contained in that line, then the sum of measurements of the two angles formed is 180°.	
Angles at a Point		Suppose three or more rays with the same vertex separate the plane into angles with disjointed interiors. Then, the sum of all the measurements of the angles is 360°.	

Angle Relationship	Definition	Diagram
Complementary Angles		
Supplementary Angles		

Lesson 1:   Complementary and Supplementary Angles

EUREKA MATH®

Exercise 1

1. In a complete sentence, describe the relevant angle relationships in the diagram. Write an equation for the angle relationship shown in the figure and solve for x. Confirm your answers by measuring the angle with a protractor.

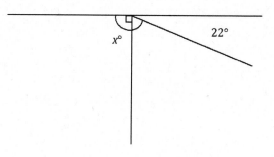

Example 1

The measures of two supplementary angles are in the ratio of $2 : 3$. Find the measurements of the two angles.

Exercises 2–4

2. In a pair of complementary angles, the measurement of the larger angle is three times that of the smaller angle. Find the measurements of the two angles.

3. The measure of a supplement of an angle is 6° more than twice the measure of the angle. Find the measurement of the two angles.

4. The measure of a complement of an angle is 32° more than three times the angle. Find the measurement of the two angles.

Example 2

Two lines meet at a point that is also the vertex of an angle. Set up and solve an appropriate equation for x and y.

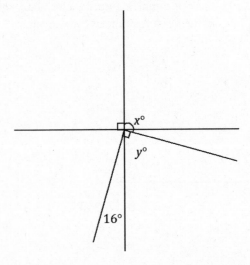

EUREKA
MATH

Lesson Summary

- Supplementary angles are two angles whose measurements sum to 180°.
- Complementary angles are two angles whose measurements sum to 90°.
- Once an angle relationship is identified, the relationship can be modeled with an equation that will find an unknown value. The unknown value may be used to find the measure of the unknown angle.

Name _____   Date _____

1. Set up and solve an equation for the value of x. Use the value of x and a relevant angle relationship in the diagram to determine the measurement of $\angle EAF$.

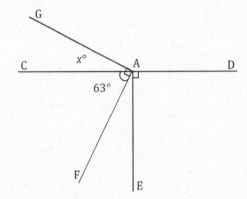

2. The measurement of the supplement of an angle is 39° more than half the angle. Find the measurement of the angle and its supplement.

1. Two lines meet at the endpoint of a ray. Set up and solve the appropriate equations to determine a and b.

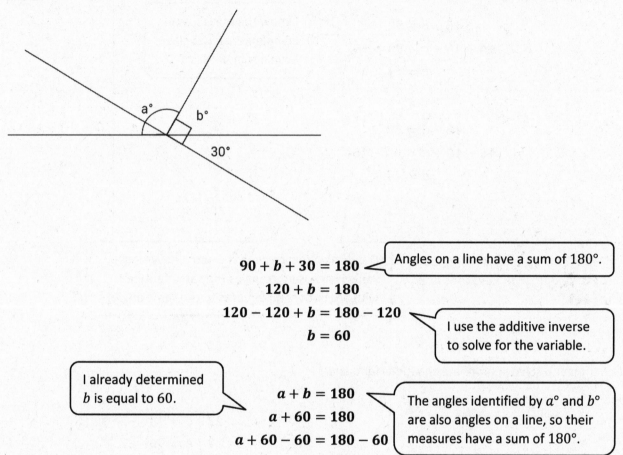

$$90 + b + 30 = 180$$

Angles on a line have a sum of 180°.

$$120 + b = 180$$
$$120 - 120 + b = 180 - 120$$
$$b = 60$$

I use the additive inverse to solve for the variable.

I already determined b is equal to 60.

$$a + b = 180$$
$$a + 60 = 180$$
$$a + 60 - 60 = 180 - 60$$
$$a = 120$$

The angles identified by $a°$ and $b°$ are also angles on a line, so their measures have a sum of 180°.

Therefore, the angles identified by $a°$ and $b°$ have measures of 120° and 60°, respectively.

EUREKA MATH

2. Two lines meet at the common endpoint of two rays. Set up and solve the appropriate equations to
determine c and d.

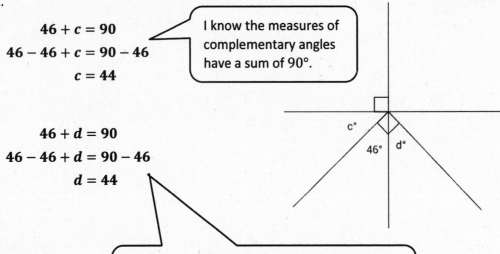

$$46 + c = 90$$
$$46 - 46 + c = 90 - 46$$
$$c = 44$$

> I know the measures of complementary angles have a sum of 90°.

$$46 + d = 90$$
$$46 - 46 + d = 90 - 46$$
$$d = 44$$

> I could also recognize that c and d have equal values because the angles they are identified with are both complements to the same angle.

Therefore, the angles identified by $c°$ and $d°$ both have measures of 44°.

3. Set up and solve appropriate equations for e and f.

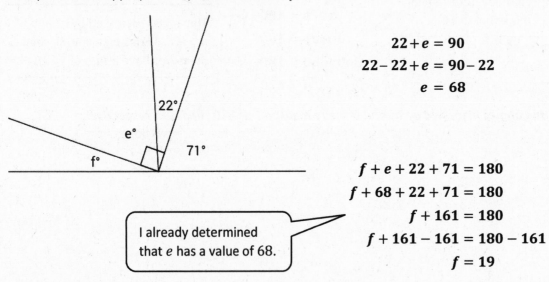

$$22 + e = 90$$
$$22 - 22 + e = 90 - 22$$
$$e = 68$$

> I already determined that e has a value of 68.

$$f + e + 22 + 71 = 180$$
$$f + 68 + 22 + 71 = 180$$
$$f + 161 = 180$$
$$f + 161 - 161 = 180 - 161$$
$$f = 19$$

Therefore, the angles identified by $e°$ and $f°$ have measures of 68° and 19°, respectively.

EUREKA MATH

4. The measurement of the supplement of an angle is 30° more than double the measurement of the angle. Find the measurements of the angle and its supplement.

Let $x°$ represent the measurement of the angle.

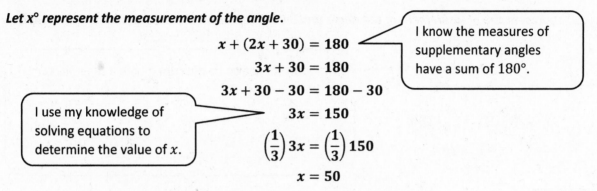

$$x + (2x + 30) = 180$$
$$3x + 30 = 180$$
$$3x + 30 - 30 = 180 - 30$$
$$3x = 150$$
$$\left(\frac{1}{3}\right)3x = \left(\frac{1}{3}\right)150$$
$$x = 50$$

I know the measures of supplementary angles have a sum of 180°.

I use my knowledge of solving equations to determine the value of x.

Let $(2x + 30)°$ represent the measurement of the supplement of the angle.

To find the measure of the supplement, I can either subtract the measure of the angle from 180°, or I can substitute the value of x into the expression $2x + 30$.

$$2x + 30$$
$$= 2(50) + 30$$
$$= 100 + 30$$
$$= 130$$

The angle measures 50°, and its supplement measures 130°.

5. The measurement of the complement of an angle exceeds the measurement of the angle by 50%. Find the measurements of the angle and its complement.

Let $x°$ represent the measurement of the angle and let $(x + 0.5x)°$ represent the measurement of its complement.

$$x + (x + 0.5x) = 90$$
$$2.5x = 90$$
$$\left(\frac{1}{2.5}\right)2.5x = \left(\frac{1}{2.5}\right)90$$
$$x = 36$$

The measure of the complement of the angle is the sum of the measure of the angle plus another 50% of its measure.

$$90 - 36 = 54$$

We could also substitute 36 into the expression $x + 0.50x$ to determine the measure of the complement.

The angle measures 36° and its complement measures 54°.

6. The ratio of the measurement of an angle to the measurement of its supplement is $2:7$. Find the measurements of the angle and its supplement.

   Let $2x°$ represent the measurement of the angle and let $7x°$ represent the measurement of its supplement.

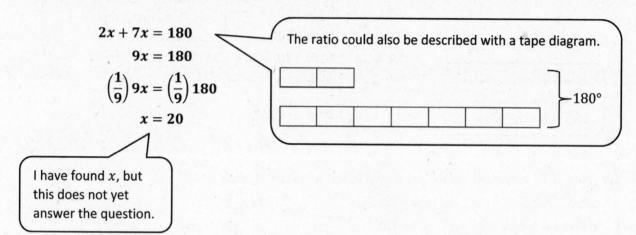

$$2x + 7x = 180$$
$$9x = 180$$
$$\left(\frac{1}{9}\right)9x = \left(\frac{1}{9}\right)180$$
$$x = 20$$

The ratio could also be described with a tape diagram.

$\}$ 180°

I have found x, but this does not yet answer the question.

Angle: $2(20)° = 40°$ **Supplement:** $7(20)° = 140°$

Therefore, the angle measures $40°$ and its supplement measures $140°$.

EUREKA
MATH

1. Two lines meet at a point that is also the endpoint of a ray. Set up and solve the appropriate equations to determine x and y.

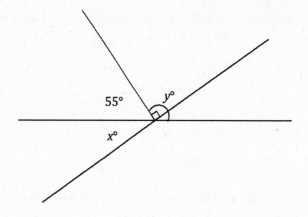

2. Two lines meet at a point that is also the vertex of an angle. Set up and solve the appropriate equations to determine x and y.

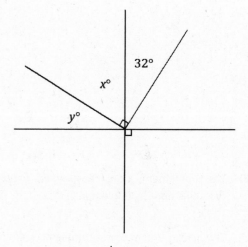

3. Two lines meet at a point that is also the vertex of an angle. Set up and solve an appropriate equation for x and y.

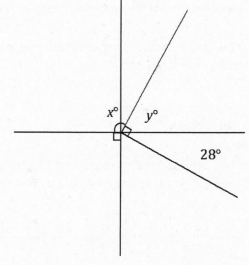

4. Set up and solve the appropriate equations for s and t.

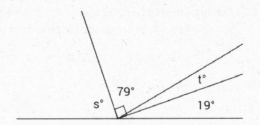

5. Two lines meet at a point that is also the endpoint of two rays. Set up and solve the appropriate equations for m and n.

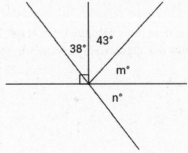

6. The supplement of the measurement of an angle is 16° less than three times the angle. Find the measurement of the angle and its supplement.

7. The measurement of the complement of an angle exceeds the measure of the angle by 25%. Find the measurement of the angle and its complement.

8. The ratio of the measurement of an angle to its complement is $1 : 2$. Find the measurement of the angle and its complement.

9. The ratio of the measurement of an angle to its supplement is $3 : 5$. Find the measurement of the angle and its supplement.

10. Let x represent the measurement of an acute angle in degrees. The ratio of the complement of x to the supplement of x is $2 : 5$. Guess and check to determine the value of x. Explain why your answer is correct.

EUREKA
MATH®

Opening Exercise

Two lines meet at a point. In a complete sentence, describe the relevant angle relationships in the diagram. Find the values of r, s, and t.

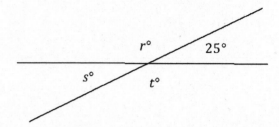

Example 1

Two lines meet at a point that is also the endpoint of a ray. In a complete sentence, describe the relevant angle relationships in the diagram. Set up and solve an equation to find the value of p and r.

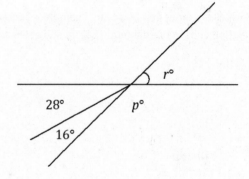

Exercise 1

Three lines meet at a point. In a complete sentence, describe the relevant angle relationship in the diagram. Set up and solve an equation to find the value of a.

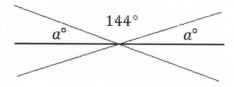

Example 2

Three lines meet at a point. In a complete sentence, describe the relevant angle relationships in the diagram. Set up and solve an equation to find the value of z.

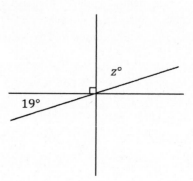

Exercise 2

Three lines meet at a point; $\angle AOF = 144°$. In a complete sentence, describe the relevant angle relationships in the diagram. Set up and solve an equation to determine the value of c.

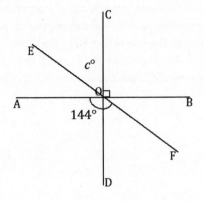

Example 3

Two lines meet at a point that is also the endpoint of a ray. The ray is perpendicular to one of the lines as shown. In a complete sentence, describe the relevant angle relationships in the diagram. Set up and solve an equation to find the value of t.

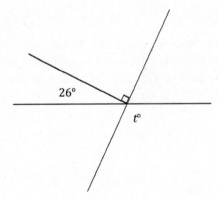

Exercise 3

Two lines meet at a point that is also the endpoint of a ray. The ray is perpendicular to one of the lines as shown. In a complete sentence, describe the relevant angle relationships in the diagram. You may add labels to the diagram to help with your description of the angle relationship. Set up and solve an equation to find the value of v.

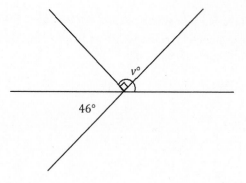

Example 4

Three lines meet at a point. In a complete sentence, describe the relevant angle relationships in the diagram. Set up and solve an equation to find the value of x. Is your answer reasonable? Explain how you know.

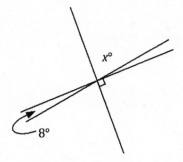

EUREKA
MATH®

Exercise 4

Two lines meet at a point that is also the endpoint of two rays. In a complete sentence, describe the relevant angle relationships in the diagram. Set up and solve an equation to find the value of x. Find the measurements of $\angle AOB$ and $\angle BOC$.

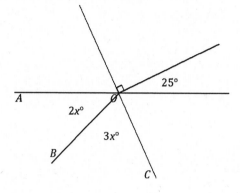

Exercise 5

a. In a complete sentence, describe the relevant angle relationships in the diagram. Set up and solve an equation to find the value of x. Find the measurements of $\angle AOB$ and $\angle BOC$.

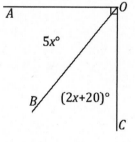

b. Katrina was solving the problem above and wrote the equation $7x + 20 = 90$. Then, she rewrote this as $7x + 20 = 70 + 20$. Why did she rewrite the equation in this way? How does this help her to find the value of x?

Lesson 2:   Solving for Unknown Angles Using Equations

EUREKA
MATH®

Lesson Summary

- To solve an unknown angle problem, identify the angle relationship(s) first to set up an equation that will yield the unknown value.

- Angles on a line and supplementary angles are not the same relationship. *Supplementary* angles are two angles whose angle measures sum to 180° whereas *angles on a line* are two or more adjacent angles whose angle measures sum to 180°.

Name _____ Date _____

Two lines meet at a point that is also the vertex of an angle. Set up and solve an equation to find the value of x. Explain why your answer is reasonable.

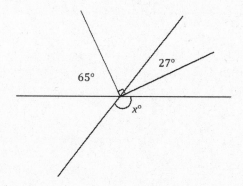

1. Two lines meet at the endpoint of a ray. Set up and solve an equation to find the values of h and j.

I know that vertical angles have the same angle measure.

$$110 = 90 + h$$
$$110 - 90 = 90 - 90 + h$$
$$20 = h$$

I know that the measures of angles on a line have a sum of 180°.

$$110 + j = 180$$
$$110 - 110 + j = 180 - 110$$
$$j = 70$$

The value of h is 20 and the value of j is 70.

2. Two lines meet at the vertex of an angle formed by two rays. Set up and solve an equation to find the value of k.

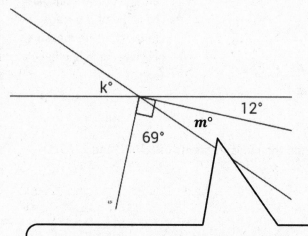

I know that the measures of complementary angles have a sum of 90°.

$$69 + m = 90$$
$$69 - 69 + m = 90 - 69$$
$$m = 21$$

I need to determine the measure of this unknown angle in order to determine the value of k. I'll let this angle have a measure of $m°$.

I know the measures of two adjacent angles that form a vertical angle to the angle $k°$.

$$21 + 12 = k$$
$$33 = k$$

The value of k is 33.

3. Three lines meet at the endpoint of a ray. Set up and solve an equation to find the value of each variable in the diagram.

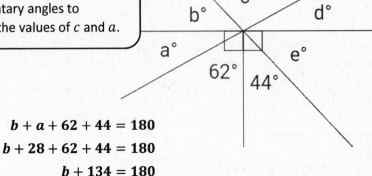

$$62 + 44 = c$$
$$106 = c$$

I can use my knowledge of vertical angles and complementary angles to determine the values of c and a.

$$a + 62 = 90$$
$$a + 62 - 62 = 90 - 62$$
$$a = 28$$

$$b + a + 62 + 44 = 180$$
$$b + 28 + 62 + 44 = 180$$
$$b + 134 = 180$$
$$b + 134 - 134 = 180 - 134$$
$$b = 46$$

Now that I know the value of a, I can calculate the value of b because I know that the measures of angles on a line have a sum of 180°.

$$b = e$$
$$46 = e$$

Now, I can use my knowledge of vertical angles to determine the values of e and d.

$$a = d$$
$$28 = d$$

4. Set up and solve an equation to find the value of x. Find the measurements of $\angle MOP$ and $\angle NOP$.

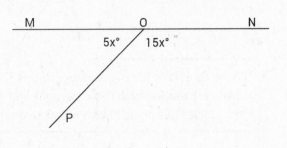

$$5x + 15x = 180$$
$$20x = 180$$
$$\left(\frac{1}{20}\right) 20x = \left(\frac{1}{20}\right) 180$$
$$x = 9$$

I can use the value of x to determine the measure of the two unknown angles.

$$\angle MOP = 5x° = 5(9)° = 45°$$
$$\angle NOP = 15x° = 15(9)° = 135°$$

I can check my answers by making sure the measures of the two angles have a sum of 180°.

EUREKA
MATH

5. Set up and solve an equation to find the value of y. Find the measurements of $\angle AOB$ and $\angle BOC$.

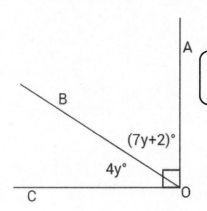

$$4y + 7y + 2 = 90$$
$$11y + 2 = 90$$
$$11y + 2 - 2 = 90 - 2$$
$$11y = 88$$
$$\left(\frac{1}{11}\right)11y = \left(\frac{1}{11}\right)88$$
$$y = 8$$

> I need to collect like terms before solving the equation.

$$\angle BOC = 4y° = 4(8)° = 32°$$
$$\angle AOB = (7y + 2)° = (7(8) + 2)° = 58°$$

> I can check my answers by making sure the measures of the two angles have a sum of 90°.

EUREKA MATH®

1. Two lines meet at a point that is also the endpoint of a ray.
   Set up and solve an equation to find the value of c.

   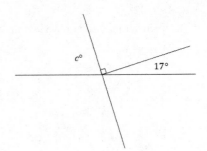

2. Two lines meet at a point that is also the endpoint of a ray. Set up and solve an equation to find the value of a. Explain why your answer is reasonable.

   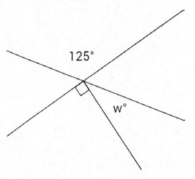

3. Two lines meet at a point that is also the endpoint of a ray. Set up and solve an equation to find the value of w.

   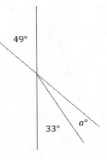

4. Two lines meet at a point that is also the vertex of an angle. Set up and solve an equation to find the value of m.

   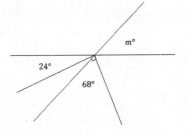

5. Three lines meet at a point. Set up and solve an equation to find the value of

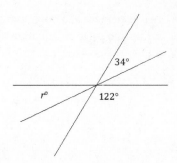

6. Three lines meet at a point that is also the endpoint of a ray. Set up and solve an equation to find the value of each variable in the diagram.

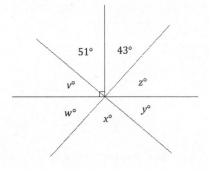

7. Set up and solve an equation to find the value of x. Find the measurement of $\angle AOB$ and of $\angle BOC$.

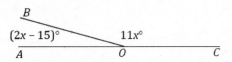

8. Set up and solve an equation to find the value of x. Find the measurement of $\angle AOB$ and of $\angle BOC$.

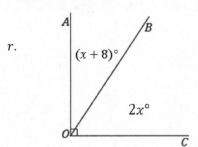

Lesson 2:   Solving for Unknown Angles Using Equations

EUREKA
MATH®

9. Set up and solve an equation to find the value of x. Find the measurement of $\angle AOB$ and of $\angle BOC$.

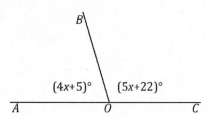

10. Write a verbal problem that models the following diagram. Then, solve for the two angles.

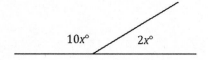

Opening Exercise

Two lines meet at a point that is also the vertex of an angle; the measurement of $\angle AOF$ is 134°. Set up and solve an equation to find the values of x and y. Are your answers reasonable? How do you know?

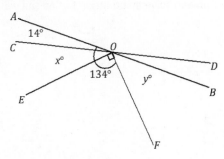

Example 1

Set up and solve an equation to find the value of x.

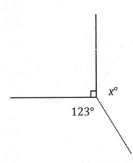

Exercise 1

Five rays meet at a common endpoint. In a complete sentence, describe the relevant angle relationships in the diagram. Set up and solve an equation to find the value of a.

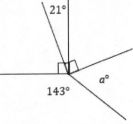

Example 2

Four rays meet at a common endpoint. In a complete sentence, describe the relevant angle relationships in the diagram. Set up and solve an equation to find the value of x. Find the measurements of $\angle BAC$ and $\angle DAE$.

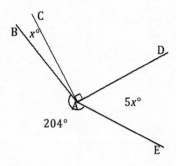

Exercise 2

Four rays meet at a common endpoint. In a complete sentence, describe the relevant angle relationships in the diagram. Set up and solve an equation to find the value of x. Find the measurement of $\angle CAD$.

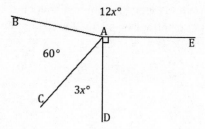

EUREKA
MATH®

Example 3

Two lines meet at a point that is also the endpoint of two rays. In a complete sentence, describe the relevant angle relationships in the diagram. Set up and solve an equation to find the value of x. Find the measurements of $\angle BAC$ and $\angle BAH$.

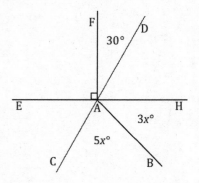

Exercise 3

Lines AB and EF meet at a point which is also the endpoint of two rays. In a complete sentence, describe the relevant angle relationships in the diagram. Set up and solve an equation to find the value of x. Find the measurements of $\angle DHF$ and $\angle AHD$.

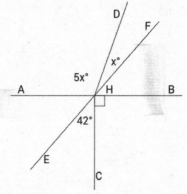

Example 4

Two lines meet at a point. Set up and solve an equation to find the value of *x*. Find the measurement of one of the vertical angles.

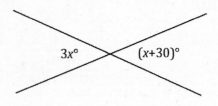

Exercise 4

Set up and solve an equation to find the value of *x*. Find the measurement of one of the vertical angles.

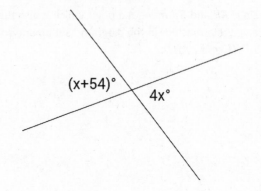

EUREKA
MATH®

Lesson Summary

Steps to Solving for Unknown Angles

- Identify the angle relationship(s).
- Set up an equation that will yield the unknown value.
- Solve the equation for the unknown value.
- Substitute the answer to determine the angle(s).
- Check and verify your answer by measuring the angle with a protractor.

Name _____   Date _____

1. Two rays have a common endpoint on a line. Set up and solve an equation to find the value of z. Find the measurements of $\angle AYC$ and $\angle DYB$.

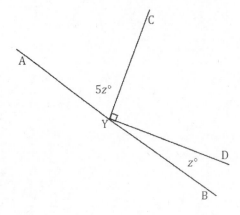

2. Two lines meet at a point that is also the vertex of an angle. Set up and solve an equation to find the value of x. Find the measurements of $\angle CAH$ and $\angle EAG$.

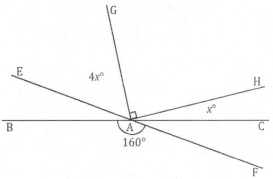

EUREKA
MATH®

1. Two lines meet at a point. Find the measurement of a vertical angle. Is your answer reasonable? Explain how you know.

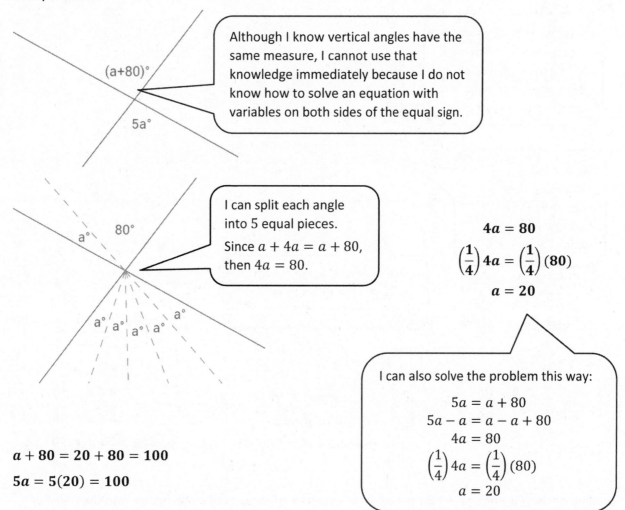

Although I know vertical angles have the same measure, I cannot use that knowledge immediately because I do not know how to solve an equation with variables on both sides of the equal sign.

I can split each angle into 5 equal pieces. Since $a + 4a = a + 80$, then $4a = 80$.

$$4a = 80$$
$$\left(\frac{1}{4}\right)4a = \left(\frac{1}{4}\right)(80)$$
$$a = 20$$

I can also solve the problem this way:
$$5a = a + 80$$
$$5a - a = a - a + 80$$
$$4a = 80$$
$$\left(\frac{1}{4}\right)4a = \left(\frac{1}{4}\right)(80)$$
$$a = 20$$

$$a + 80 = 20 + 80 = 100$$
$$5a = 5(20) = 100$$

Therefore, each vertical angle has a measure of $100°$.

My answer is reasonable because the vertical angle looks to be close to the measure of a right angle.

EUREKA MATH®

2.   Three lines meet at the endpoint of a ray. Set up and solve an equation to find the value of b.

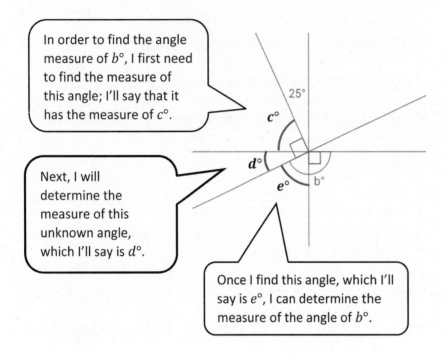

In order to find the angle measure of $b°$, I first need to find the measure of this angle; I'll say that it has the measure of $c°$.

Next, I will determine the measure of this unknown angle, which I'll say is $d°$.

Once I find this angle, which I'll say is $e°$, I can determine the measure of the angle of $b°$.

$$c + 25 = 90$$
$$c + 25 - 25 = 90 - 25$$
$$c = 65$$

$$c + d = 90$$
$$65 + d = 90$$
$$65 - 65 + d = 90 - 65$$
$$d = 25$$

$$d + e = 90$$
$$25 + e = 90$$
$$25 - 25 + e = 90 - 25$$
$$e = 65$$

Looking at the diagram, I see that the angle of measure $b°$ consists of a right angle and the angle of measure $e°$.

$90 + 65 = 155$

Therefore, the value of b is 155.

EUREKA
MATH®

3. Four angles meet at a point. The second angle measures 10° more than the first angle, the third angle measures 15° more than the second angle, and the fourth angle measures 20° more than the third angle. Find the measurements of all four angles.

> I know that the measures of angles that meet at a point have a sum of 360°.

Let x represent the value of the first angle measurement.

$$(x) + (x + 10) + (x + 10 + 15) + (x + 10 + 15 + 20) = 360$$

> Each set of parentheses represents the measure of each of the angles.

$$4x + 80 = 360$$
$$4x + 80 - 80 = 360 - 80$$
$$4x = 280$$
$$\left(\frac{1}{4}\right)4x = \left(\frac{1}{4}\right)280$$
$$x = 70$$

The following are the measures of each of the angles:

Angle 1: $70°$

Angle 2: $(70)° + 10° = 80°$

Angle 3: $(70)° + 10° + 15° = 95°$

Angle 3: $(70)° + 10° + 15° + 20° = 115°$

> To check my answers, I could add the measures of the four angles together to determine if they have a sum of 360°.

1. Two lines meet at a point. Set up and solve an equation to find the value of x.

2. Three lines meet at a point. Set up and solve an equation to find the value of a. Is your answer reasonable? Explain how you know.

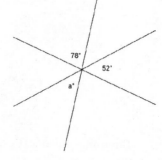

3. Two lines meet at a point that is also the endpoint of two rays. Set up and solve an equation to find the values of a and b.

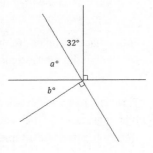

4. Three lines meet at a point that is also the endpoint of a ray. Set up and solve an equation to find the values of x and y.

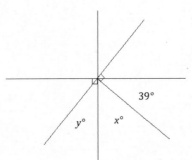

5. Two lines meet at a point. Find the measurement of one of the vertical angles. Is your answer reasonable? Explain how you know.

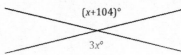

6. Three lines meet at a point that is also the endpoint of a ray. Set up and solve an equation to find the value of y.

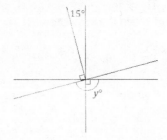

7. Three adjacent angles are at a point. The second angle is $20°$ more than the first, and the third angle is $20°$ more than the second angle.

   a. Find the measurements of all three angles.

   b. Compare the expressions you used for the three angles and their combined expression. Explain how they are equal and how they reveal different information about this situation.

8. Four adjacent angles are on a line. The measurements of the four angles are four consecutive even numbers. Determine the measurements of all four angles.

9. Three adjacent angles are at a point. The ratio of the measurement of the second angle to the measurement of the first angle is $4:3$. The ratio of the measurement of the third angle to the measurement of the second angle is $5:4$. Determine the measurements of all three angles.

10. Four lines meet at a point. Solve for x and y in the following diagram.

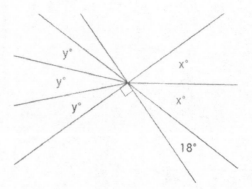

EUREKA
MATH®

Opening Exercise

The complement of an angle is four times the measurement of the angle. Find the measurement of the angle and its complement.

Find the measurements of $\angle FAE$ and $\angle CAD$.

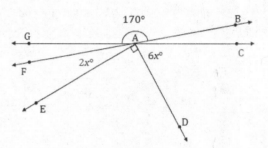

Two lines meet at a point. List the relevant angle relationship in the diagram. Set up and solve an equation to find the value of x. Find the measurement of one of the vertical angles.

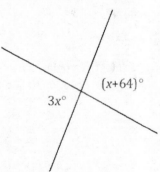

Exercise 1

Set up and solve an equation to find the value of x. List the relevant angle relationship in the diagram. Find the measurement of one of the vertical angles.

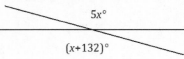

Example 2

Three lines meet at a point. List the relevant angle relationships in the diagram. Set up and solve an equation to find the value of b.

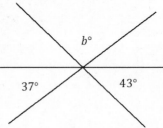

Exercise 2

Two lines meet at a point that is also the endpoint of two rays. List the relevant angle relationships in the diagram. Set up and solve an equation to find the value of b.

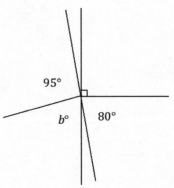

EUREKA MATH

Example 3

The measurement of an angle is $\frac{2}{3}$ the measurement of its supplement. Find the measurements of the angle and its supplement.

Exercise 3

The measurement of an angle is $\frac{1}{4}$ the measurement of its complement. Find the measurements of the two complementary angles.

Example 4

Three lines meet at a point that is also the endpoint of a ray. List the relevant angle relationships in the diagram. Set up and solve an equation to find the value of z.

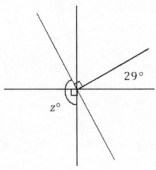

Exercise 4

Two lines meet at a point that is also the vertex of an angle. Set up and solve an equation to find the value of x. Find the measurements of $\angle GAF$ and $\angle BAC$.

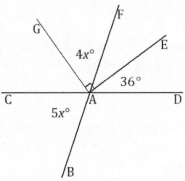

EUREKA
MATH®

Lesson Summary

Steps to Solving for Unknown Angles

- Identify the angle relationship(s).
- Set up an equation that will yield the unknown value.
- Solve the equation for the unknown value.
- Substitute the answer to determine the measurement of the angle(s).
- Check and verify your answer by measuring the angle with a protractor.

Name _____ Date _____

Lines BC and EF meet at A. Rays AG and AD form a right angle. Set up and solve an equation to find the values of x and w.

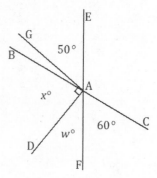

EUREKA
MATH®

1. $\overrightarrow{BE}$ and $\overrightarrow{AD}$ meet at G. Set up and solve an equation to find the value of y. Find the measurements of $\angle BGC$ and $\angle CGD$.

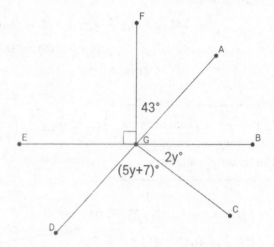

$$43 + \angle AGB = 90$$
$$43 - 43 + \angle AGB = 90 - 43$$
$$\angle AGB = 47$$

> I need to determine the measurement of $\angle AGB$ before finding the measurements of $\angle BGC$ and $\angle CGD$.

$$\angle AGB + 2y + 5y + 7 = 180$$
$$47 + 2y + 5y + 7 = 180$$
$$54 + 7y = 180$$
$$54 - 54 + 7y = 180 - 54$$
$$7y = 126$$
$$y = 18$$

> Now that I know the value of y, I can calculate the measurements of $\angle BGC$ and $\angle CGD$.

The measurement of $\angle BGC$: $2(18)° = 36°$

The measurement of $\angle CGD$: $(5(18) + 7)° = (90 + 7)° = 97°$

2. Five rays meet at a point. Set up and solve an equation to find the value of m. Find the measurements of $\angle EDF$ and $\angle HDG$.

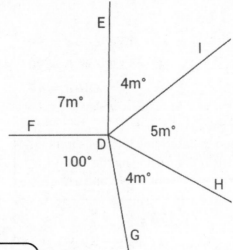

$$7m + 4m + 5m + 4m + 100 = 360$$
$$20m + 100 = 360$$
$$20m + 100 - 100 = 360 - 100$$
$$20m = 260$$
$$\left(\frac{1}{20}\right)20m = \left(\frac{1}{20}\right)260$$
$$m = 13$$

> I can use the value of m to determine the measures of the unknown angles.

$$\angle EDF = 7m° = 7(13)° = 91°$$
$$\angle HDG = 4m° = 4(13)° = 52°$$

> I only need to determine the measures of the two angles presented in the question.

3. Three adjacent angles form a line. The measurement of each angle is one of three consecutive, positive whole numbers. Determine the measurements of all three angles.

> Consecutive numbers are ones that directly follow each other. For example, 2, 3, 4.

Let $x°$ represent the smallest angle measure.

> Since $x°$ represents the measure of the smallest angle, then the measure of the second angle is 1° larger than $x°$, and the measure of the third angle is 2° larger than $x°$.

$$x + (x + 1) + (x + 2) = 180$$
$$3x + 3 = 180$$
$$3x + 3 - 3 = 180 - 3$$
$$3x = 177$$
$$\left(\frac{1}{3}\right)3x = \left(\frac{1}{3}\right)177$$
$$x = 59$$

The three angles measure $59°$, $60°$, and $61°$ because I determined the smallest angle measures $59°$, and the measures of the other two angles are consecutive numbers.

EUREKA MATH®

4. The ratio of measurement of an angle to the measurement of its supplement is $1:4$.

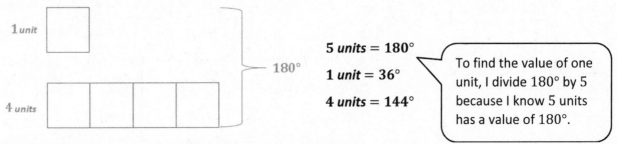

5 units = 180°

1 unit = 36°

4 units = 144°

To find the value of one unit, I divide 180° by 5 because I know 5 units has a value of 180°.

The measure of the angle that satisfies these criteria is 36°.

5. The sum of four times the measurement of the complement of an angle and the measurement of the supplement of that angle is 240°. What is the measurement of the angle?

Let $a°$ represent the measurement of the angle.

I use my knowledge of complements and supplements to write an equation.

$$4(90 - a) + (180 - a) = 240$$
$$360 - 4a + 180 - a = 240$$
$$540 - 5a = 240$$
$$540 - 540 - 5a = 240 - 540$$
$$-5a = -300$$
$$\left(-\frac{1}{5}\right)(-5a) = \left(-\frac{1}{5}\right)(-300)$$
$$a = 60$$

I need to be careful when collecting like terms. I remember that subtracting is the same as adding the opposite, so I could rewrite the equation as follows:
$360 + (-4a) + 180 + (-a) = 240$.

The measurement of the angle is 60°.

EUREKA
MATH

1. Four rays have a common endpoint on a line. Set up and solve an equation to find the value of c.

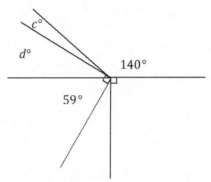

2. Lines BC and EF meet at A. Set up and solve an equation to find the value of x. Find the measurements of $\angle EAH$ and $\angle HAC$.

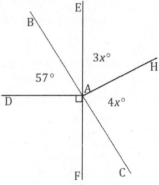

3. Five rays share a common endpoint. Set up and solve an equation to find the value of x. Find the measurements of $\angle DAG$ and $\angle GAH$.

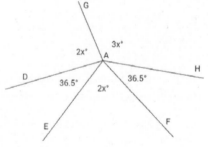

4. Four lines meet at a point which is also the endpoint of three rays. Set up and solve an equation to find the values of x and y.

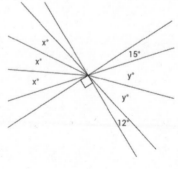

5. Two lines meet at a point that is also the vertex of a right angle. Set up and solve an equation to find the value of x. Find the measurements of $\angle CAE$ and $\angle BAG$.

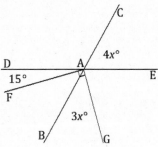

6. Five angles are at a point. The measurement of each angle is one of five consecutive, positive whole numbers.

   a. Determine the measurements of all five angles.

   b. Compare the expressions you used for the five angles and their combined expression. Explain how they are equivalent and how they reveal different information about this situation.

7. Let $x°$ be the measurement of an angle. The ratio of the measurement of the complement of the angle to the measurement of the supplement of the angle is $1 : 3$. The measurement of the complement of the angle and the measurement of the supplement of the angle have a sum of $180°$. Use a tape diagram to find the measurement of this angle.

8. Two lines meet at a point. Set up and solve an equation to find the value of x. Find the measurement of one of the vertical angles.

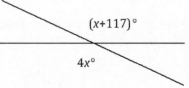

9. The difference between three times the measurement of the complement of an angle and the measurement of the supplement of that angle is $20°$. What is the measurement of the angle?

© 2019 Great Minds®. eureka-math.org

EUREKA
MATH

Opening

When studying triangles, it is essential to be able to communicate about the parts of a triangle without any confusion. The following terms are used to identify particular angles or sides:

- between
- adjacent to
- opposite to
- included [side/angle]

Exercises 1–7

Use the figure $\triangle ABC$ to fill in the following blanks.

1. $\angle A$ is _____ sides $\overline{AB}$ and $\overline{AC}$.

2. $\angle B$ is _____ side $\overline{AB}$ and to side $\overline{BC}$.

3. Side $\overline{AB}$ is _____ $\angle C$.

4. Side _____ is the included side of $\angle B$ and $\angle C$.

5. $\angle$ _____ is opposite to side $\overline{AC}$.

6. Side $\overline{AB}$ is between $\angle$_____ and $\angle$_____.

7. What is the included angle of sides $\overline{AB}$ and $\overline{BC}$ _____

Now that we know what to call the parts within a triangle, we consider how to discuss two triangles. We need to compare the parts of the triangles in a way that is easy to understand. To establish some alignment between the triangles, we pair up the vertices of the two triangles. We call this a *correspondence*. Specifically, a correspondence between two triangles is a pairing of each vertex of one triangle with one (and only one) vertex of the other triangle. A correspondence provides a systematic way to compare parts of two triangles.

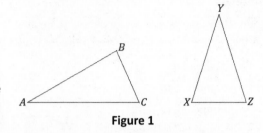

Figure 1

In Figure 1, we can choose to assign a correspondence so that A matches to X, B matches to Y, and C matches to Z. We notate this correspondence with double arrows: $A \leftrightarrow X$, $B \leftrightarrow Y$, and $C \leftrightarrow Z$. This is just one of six possible correspondences between the two triangles. Four of the six correspondences are listed below; find the remaining two correspondences.

$$\begin{array}{ll}
A \longleftrightarrow X & A \longleftrightarrow X \\
B \longleftrightarrow Y & B \ \times\ Y \\
C \longleftrightarrow Z & C \qquad Z \\
A \ \times\ X & A \qquad X \\
B \qquad Y & B \ \times\ Y \\
C \longleftrightarrow Z & C \qquad Z
\end{array}$$

A simpler way to indicate the triangle correspondences is to let the order of the vertices define the correspondence (i.e., the first corresponds to the first, the second to the second, and the third to the third). The correspondences above can be written in this manner. Write the remaining two correspondences in this way.

$$\triangle\, ABC \leftrightarrow \triangle\, XYZ \qquad \triangle\, ABC \leftrightarrow \triangle\, XZY$$

$$\triangle\, ABC \leftrightarrow \triangle\, YXZ \qquad \triangle\, ABC \leftrightarrow \triangle\, YZX$$

With a correspondence in place, comparisons can be made about corresponding sides and corresponding angles. The following are corresponding vertices, angles, and sides for the triangle correspondence $\triangle\, ABC \leftrightarrow \triangle\, YXZ$. Complete the missing correspondences.

Vertices:	$A \leftrightarrow Y$	$B \leftrightarrow$	$C \leftrightarrow$
Angles:	$\angle A \leftrightarrow \angle Y$	$\angle B \leftrightarrow$	$\angle C \leftrightarrow$
Sides:	$\overline{AB} \leftrightarrow \overline{YX}$	$\overline{BC} \leftrightarrow$	$\overline{CA} \leftrightarrow$

EUREKA MATH

Example 1

Given the following triangle correspondences, use double arrows to show the correspondence between vertices, angles, and sides.

Triangle Correspondence	$\triangle ABC \leftrightarrow \triangle STR$
Correspondence of Vertices	
Correspondence of Angles	
Correspondence of Sides	

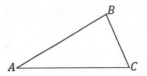

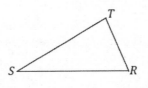

Examine Figure 2. By simply looking, it is impossible to tell the two triangles apart unless they are labeled. They look exactly the same (just as identical twins look the same). One triangle could be picked up and placed on top of the other.

Two triangles are identical if there is a triangle correspondence so that corresponding sides and angles of each triangle are equal in measurement. In Figure 2, there is a correspondence that will match up equal sides and equal angles, $\triangle ABC \leftrightarrow \triangle XYZ$; we can conclude that $\triangle ABC$ is identical to $\triangle XYZ$. This is not to say that we cannot find a correspondence in Figure 2 so that unequal sides and unequal angles are matched up, but there certainly is one correspondence that will match up angles with equal measurements and sides of equal lengths, making the triangles identical.

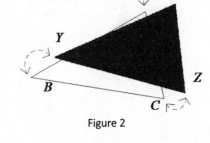

Figure 2

In discussing identical triangles, it is useful to have a way to indicate those sides and angles that are equal. We mark sides with tick marks and angles with arcs if we want to draw attention to them. If two angles or two sides have the same number of marks, it means they are equal.

In this figure, $AC = DE = EF$, and $\angle B = \angle E$.

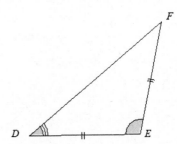

Example 2

Two identical triangles are shown below. Give a triangle correspondence that matches equal sides and equal angles.

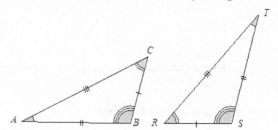

Exercise 8

8. Sketch two triangles that have a correspondence. Describe the correspondence in symbols or words. Have a partner check your work.

EUREKA
MATH

Lesson Summary

- Two triangles and their respective parts can be compared once a correspondence has been assigned to the two triangles. Once a correspondence is selected, corresponding sides and corresponding angles can also be determined.

- Double arrows notate corresponding vertices. Triangle correspondences can also be notated with double arrows.

- Triangles are identical if there is a correspondence so that corresponding sides and angles are equal.

- An equal number of tick marks on two different sides indicates the sides are equal in measurement. An equal number of arcs on two different angles indicates the angles are equal in measurement.

Name _____   Date _____

1. The following triangles are identical and have the correspondence $\triangle ABC \leftrightarrow \triangle YZX$. Find the measurements for each of the following sides and angles. Figures are not drawn to scale.

$AB =$ _____

_____ $= ZX$

_____ $= XY$

$\angle A =$ _____

$\angle B =$ _____

_____ $= \angle X$

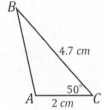
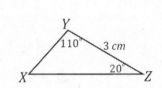

2. Explain why correspondences are useful.

1. Given the following triangles' correspondences, use double arrows to show the correspondence between vertices, angles, and sides.

> The order in which the vertices are listed for each triangle is important. The correspondence of vertices, angles, and sides are determined by the order of the label.

> The first vertex labeled in the first triangle corresponds with the first vertex labeled in the second triangle. The same is true for the other two vertices.

> The same correspondence rule for vertices is true for angle correspondence.

Triangle Correspondence	$\triangle MNP \leftrightarrow \triangle XYZ$
Correspondence of Vertices	$M \longleftrightarrow X$ $N \longleftrightarrow Y$ $P \longleftrightarrow Z$
Correspondence of Angles	$\angle M \longleftrightarrow \angle X$ $\angle N \longleftrightarrow \angle Y$ $\angle P \longleftrightarrow \angle Z$
Correspondence of Sides	$\overline{MN} \longleftrightarrow \overline{XY}$ $\overline{NP} \longleftrightarrow \overline{YZ}$ $\overline{MP} \longleftrightarrow \overline{XZ}$

> The sides of triangles are line segments and are defined by two vertices. The corresponding sides also depend on the order of the triangle labels.

2. Name the angle pairs and side pairs to find a triangle correspondence that matches sides of equal length and angles of equal measurement.

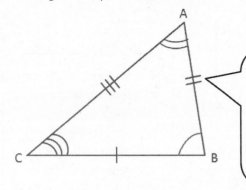
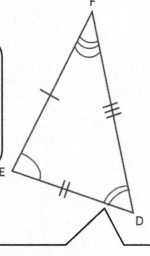

The tick marks on the triangles' sides indicate correspondence. Corresponding sides have the same number of tick marks.

$$AB = \boldsymbol{DE} \qquad BC = \boldsymbol{EF} \qquad AC = \boldsymbol{DF}$$

$$\angle A = \angle \boldsymbol{D} \qquad \angle B = \angle \boldsymbol{E} \qquad \angle C = \angle \boldsymbol{F}$$

$$\triangle ABC \leftrightarrow \triangle \boldsymbol{DEF}$$

The arcs in the triangles' angles indicate correspondence. Corresponding angles have the same number of arcs.

I know the order that I use to name the triangles is important. The letters of the corresponding angles must be in the same position for both triangles.

EUREKA MATH

Given the following triangle correspondences, use double arrows to show the correspondence between vertices, angles, and sides.

1.

Triangle Correspondence	△ *ABC* ↔ △ *RTS*
Correspondence of Vertices	
Correspondence of Angles	
Correspondence of Sides	

2.

Triangle Correspondence	△ *ABC* ↔ △ *FGE*
Correspondence of Vertices	
Correspondence of Angles	
Correspondence of Sides	

3.

Triangle Correspondence	△ *QRP* ↔ △ *WYX*
Correspondence of Vertices	
Correspondence of Angles	
Correspondence of Sides	

Name the angle pairs and side pairs to find a triangle correspondence that matches sides of equal length and angles of equal measurement.

4.

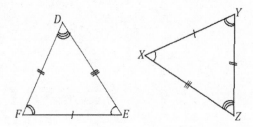

5.

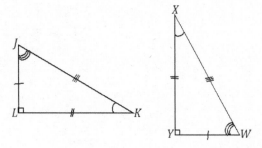

6.

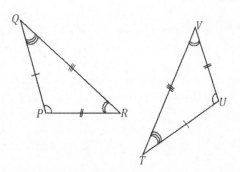

7. Consider the following points in the coordinate plane.

a. How many different (non-identical) triangles can be drawn using any three of these six points as vertices?

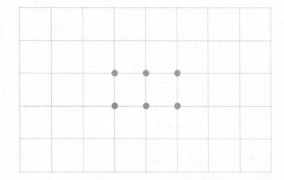

b. How can we be sure that there are no more possible triangles?

EUREKA
MATH®

8. Quadrilateral $ABCD$ is identical with quadrilateral $WXYZ$ with a correspondence $A \leftrightarrow W$, $B \leftrightarrow X$, $C \leftrightarrow Y$, and $D \leftrightarrow Z$.

    a. In the figure below, label points W, X, Y, and Z on the second quadrilateral.

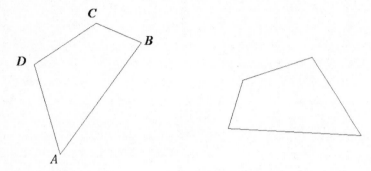

    b. Set up a correspondence between the side lengths of the two quadrilaterals that matches sides of equal length.

    c. Set up a correspondence between the angles of the two quadrilaterals that matches angles of equal measure.

EUREKA
MATH®

Exploratory Challenge

Use a ruler, protractor, and compass to complete the following problems.

1. Use your ruler to draw three segments of the following lengths: 4 cm, 7.2 cm, and 12.8 cm. Label each segment with its measurement.

2. Draw complementary angles so that one angle is 35°. Label each angle with its measurement. Are the angles required to be adjacent?

3. Draw vertical angles so that one angle is 125°. Label each angle formed with its measurement.

4. Draw three distinct segments of lengths 2 cm, 4 cm, and 6 cm. Use your compass to draw three circles, each with a radius of one of the drawn segments. Label each radius with its measurement.

5. Draw three adjacent angles a, b, and c so that $a = 25°$, $b = 90°$, and $c = 50°$. Label each angle with its measurement.

6. Draw a rectangle $ABCD$ so that $AB = CD = 8$ cm and $BC = AD = 3$ cm.

Lesson 6:   Drawing Geometric Shapes

EUREKA MATH

7. Draw a segment AB that is 5 cm in length. Draw a second segment that is longer than $\overline{AB}$, and label one endpoint C. Use your compass to find a point on your second segment, which will be labeled D, so that $CD = AB$.

8. Draw a segment AB with a length of your choice. Use your compass to construct two circles:

    i.   A circle with center A and radius AB.

    ii. A circle with center B and radius BA.

    Describe the construction in a sentence.

9. Draw a horizontal segment AB, 12 cm in length.

    a. Label a point O on $\overline{AB}$ that is 4 cm from B.

    b. Point O will be the vertex of an angle COB.

    c. Draw ray OC so that the ray is above $\overline{AB}$ and $\angle COB = 30°$.

    d. Draw a point P on $\overline{AB}$ that is 4 cm from A.

    e. Point P will be the vertex of an angle QPO.

    f. Draw ray PQ so that the ray is above $\overline{AB}$ and $\angle QPO = 30°$.

10. Draw segment AB of length 4 cm. Draw two circles that are the same size, one with center A and one with center B (i.e., do not adjust your compass in between) with a radius of a length that allows the two circles to intersect in two distinct locations. Label the points where the two circles intersect C and D. Join A and C with a segment; join B and C with a segment. Join A and D with a segment; join B and D with a segment.

    What kind of triangles are $\triangle ABC$ and $\triangle ABD$? Justify your response.

EUREKA
MATH®

11. Determine all possible measurements in the following triangle, and use your tools to create a copy of it.

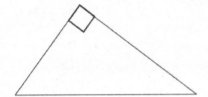

Lesson Summary

The compass is a tool that can be used for many purposes that include the following:

- Constructing circles.
- Measuring and marking a segment of equal length to another segment.
- Confirming that the radius of the center of a circle to the circle itself remains constant no matter where you are on the circle.

Name _____ Date _____

1. Draw a square *PQRS* with side length equal to 5 cm. Label the side and angle measurements.

2. Draw a segment *AB*, 6 cm in length. Draw a circle whose diameter is segment *AB*.

Necessary Tools

Students need a ruler, protractor, and compass to complete the homework assignment.

> I use rulers to measure and draw line segments, protractors to construct angles, and compasses to draw circles.

Use a ruler, protractor, and compass to complete the following problems.

1. Draw a segment BC that is 4 cm in length, perpendicular to segment DE, which is 7 cm in length.

> I know that perpendicular tells me that the two segments will create a right angle.

> I first use my ruler to draw segment BC.

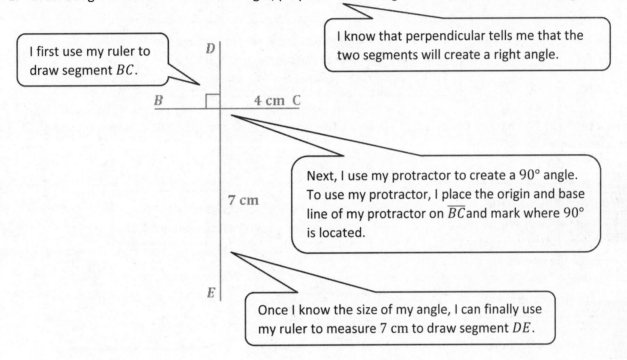

> Next, I use my protractor to create a 90° angle. To use my protractor, I place the origin and base line of my protractor on $\overline{BC}$ and mark where 90° is located.

> Once I know the size of my angle, I can finally use my ruler to measure 7 cm to draw segment DE.

2. Draw △ XYZ so that ∠Y has a measurement of 75°.

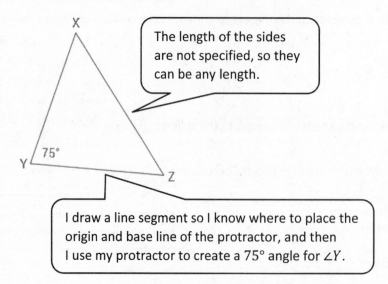

The length of the sides are not specified, so they can be any length.

I draw a line segment so I know where to place the origin and base line of the protractor, and then I use my protractor to create a 75° angle for ∠Y.

3. Draw a segment BC that is 2 cm in length. Draw a circle with center B and radius BC. Draw a second circle with diameter BC.

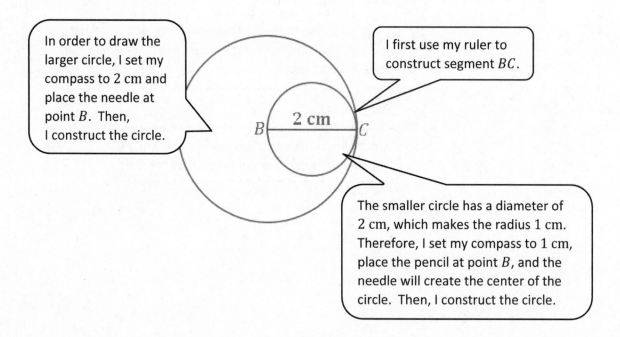

In order to draw the larger circle, I set my compass to 2 cm and place the needle at point B. Then, I construct the circle.

I first use my ruler to construct segment BC.

The smaller circle has a diameter of 2 cm, which makes the radius 1 cm. Therefore, I set my compass to 1 cm, place the pencil at point B, and the needle will create the center of the circle. Then, I construct the circle.

Lesson 6:   Drawing Geometric Shapes

EUREKA MATH

Use a ruler, protractor, and compass to complete the following problems.

1. Draw a segment AB that is 5 cm in length and perpendicular to segment CD, which is 2 cm in length.

2. Draw supplementary angles so that one angle is 26°. Label each angle with its measurement.

3. Draw △ ABC so that ∠B has a measurement of 100°.

4. Draw a segment AB that is 3 cm in length. Draw a circle with center A and radius AB. Draw a second circle with diameter AB.

5. Draw an isosceles △ ABC. Begin by drawing ∠A with a measurement of 80°. Use the rays of ∠A as the equal legs of the triangle. Choose a length of your choice for the legs, and use your compass to mark off each leg. Label each marked point with B and C. Label all angle measurements.

6. Draw an isosceles △ DEF. Begin by drawing a horizontal segment DE that is 6 cm in length. Use your protractor to draw ∠D and ∠E so that the measurements of both angles are 30°. If the non-horizontal rays of ∠D and ∠E do not already cross, extend each ray until the two rays intersect. Label the point of intersection F. Label all side and angle measurements.

7. Draw a segment AB that is 7 cm in length. Draw a circle with center A and a circle with center B so that the circles are not the same size, but do intersect in two distinct locations. Label one of these intersections C. Join A to C and B to C to form △ ABC.

8. Draw an isosceles trapezoid $WXYZ$ with two equal base angles, ∠W and ∠X, that each measures 110°. Use your compass to create the two equal sides of the trapezoid. Leave arc marks as evidence of the use of your compass. Label all angle measurements. Explain how you constructed the trapezoid.

Example 1

Use what you know about drawing parallel lines with a setsquare to draw rectangle $ABCD$ with dimensions of your choice. State the steps you used to draw your rectangle, and compare those steps to those of a partner.

Example 2

Use what you know about drawing parallel lines with a setsquare to draw rectangle $ABCD$ with $AB = 3$ cm and $BC = 5$ cm. Write a plan for the steps you will take to draw $ABCD$.

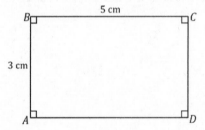

Example 3

Use a setsquare, ruler, and protractor to draw parallelogram $PQRS$ so that the measurement of $\angle P$ is 50°, $PQ = 5$ cm, the measurement of $\angle Q$ is 130°, and the length of the altitude to $\overline{PQ}$ is 4 cm.

Exercise 1

Use a setsquare, ruler, and protractor to draw parallelogram $DEFG$ so that the measurement of $\angle D$ is 40°, $DE = 3$ cm, the measurement of $\angle E$ is 140°, and the length of the altitude to $\overline{DE}$ is 5 cm.

Lesson 7:   Drawing Parallelograms

EUREKA MATH

Example 4

Use a setsquare, ruler, and protractor to draw rhombus $ABCD$ so that the measurement of $\angle A = 80°$, the measurement of $\angle B = 100°$, and each side of the rhombus measures 5 cm.

EUREKA
MATH®

Lesson Summary

A protractor, ruler, and setsquare are necessary tools to construct a parallelogram. A setsquare is the tool that gives a means to draw parallel lines for the sides of a parallelogram.

Name _____   Date _____

Use what you know about drawing parallel lines with a setsquare to draw square $ABCD$ with $AB = 5$ cm. Explain how you created your drawing.

Necessary Tools

Students need a ruler, protractor, and setsquare to complete this homework assignment. Students created setsquares at school, but, if necessary, students can follow the diagrams below to make a new one.

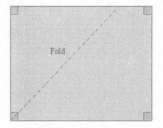 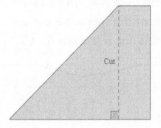 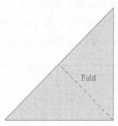 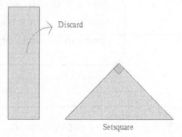

1. Use a setsquare and a ruler to construct rectangle $WXYZ$ with $WX = 4$ cm and $XY = 5$ cm.

First, I draw segment WX by using my ruler to measure 4 cm.

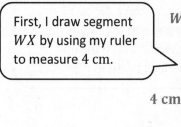

Finally, I need to create right angles and connect the two parallel segments. To do this, I align one leg of my setsquare with $\overline{WX}$, line up my ruler so the outer portion goes through point X, and then I draw a segment. I mark the point that intersects the segment parallel to $\overline{WX}$ as Y. I repeat this process to find point Z.

Second, I use the setsquare to create a segment that is parallel to $\overline{WX}$. To do this, I align one leg of the setsquare with $\overline{WX}$ and place the ruler along the other leg of the setsquare. I make a mark 5 cm away from $\overline{WX}$. I slide the setsquare along the ruler until I reach the mark and use the leg of the setsquare to draw a segment through the mark that is parallel to $\overline{WX}$.

2. Use a setsquare, ruler, and protractor to draw parallelogram $ABCD$ so that $\angle A = 60°$, $AB = 4$ cm, $\angle B = 120°$, and the altitude to $\overline{AB}$ is 7 cm.

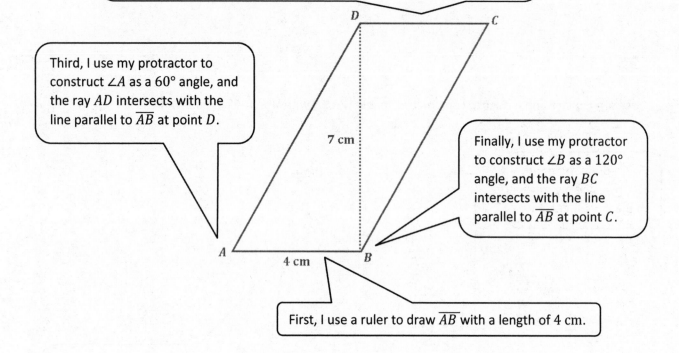

Second, I align the setsquare and ruler so that one leg of the setsquare aligns with $\overline{AB}$, and I mark a point, X, 7 cm from $\overline{AB}$. Then, I slide the setsquare along the ruler so that one side of the setsquare passes through X, and I draw a line through X; this line is parallel to $\overline{AB}$.

Third, I use my protractor to construct $\angle A$ as a 60° angle, and the ray AD intersects with the line parallel to $\overline{AB}$ at point D.

Finally, I use my protractor to construct $\angle B$ as a 120° angle, and the ray BC intersects with the line parallel to $\overline{AB}$ at point C.

First, I use a ruler to draw $\overline{AB}$ with a length of 4 cm.

7 cm

4 cm

Lesson 7:   Drawing Parallelograms

EUREKA MATH

1. Draw rectangle $ABCD$ with $AB = 5$ cm and $BC = 7$ cm.

2. Use a setsquare, ruler, and protractor to draw parallelogram $PQRS$ so that the measurement of $\angle P$ is 65°, $PQ = 8$ cm, the measurement of $\angle Q$ is 115°, and the length of the altitude to $\overline{PQ}$ is 3 cm.

3. Use a setsquare, ruler, and protractor to draw rhombus $ABCD$ so that the measurement of $\angle A$ is 60°, and each side of the rhombus measures 5 cm.

The following table contains partial information for parallelogram $ABCD$. Using no tools, make a sketch of the parallelogram. Then, use a ruler, protractor, and setsquare to draw an accurate picture. Finally, complete the table with the unknown lengths.

	$\angle A$	AB	Altitude to $\overline{AB}$	BC	Altitude to $\overline{BC}$
4.	45°	5 cm		4 cm	
5.	50°	3 cm		3 cm	
6.	60°	4 cm	4 cm		

7. Use what you know about drawing parallel lines with a setsquare to draw trapezoid $ABCD$ with parallel sides $\overline{AB}$ and $\overline{CD}$. The length of $\overline{AB}$ is 3 cm, and the length of $\overline{CD}$ is 5 cm; the height between the parallel sides is 4 cm. Write a plan for the steps you will take to draw $ABCD$.

8. Use the appropriate tools to draw rectangle $FIND$ with $FI = 5$ cm and $IN = 10$ cm.

9. Challenge: Determine the area of the largest rectangle that will fit inside an equilateral triangle with side length 5 cm.

Exercises 1–2

1. Use your protractor and ruler to draw right triangle DEF. Label all sides and angle measurements.

    a. Predict how many of the right triangles drawn in class are identical to the triangle you have drawn.

    b. How many of the right triangles drawn in class are identical to the triangle you drew? Were you correct in your prediction?

2. Given the following three sides of $\triangle ABC$, use your compass to copy the triangle. The longest side has been copied for you already. Label the new triangle $A'B'C'$, and indicate all side and angle measurements. For a reminder of how to begin, refer to Lesson 6 Exploratory Challenge Problem 10.

    A_____B

    B_____C

    A_____C

    A_____C

Exploratory Challenge

A triangle is to be drawn provided the following conditions: the measurements of two angles are 30° and 60°, and the length of a side is 10 cm. Note that where each of these measurements is positioned is not fixed.

a. How is the premise of this problem different from Exercise 2?

b. Given these measurements, do you think it will be possible to draw more than one triangle so that the triangles drawn will be different from each other? Or do you think attempting to draw more than one triangle with these measurements will keep producing the same triangle, just turned around or flipped about?

c. Based on the provided measurements, draw $\triangle ABC$ so that $\angle A = 30°$, $\angle B = 60°$, and $AB = 10$ cm. Describe how the 10 cm side is positioned.

Lesson 8: Drawing Triangles

EUREKA MATH

d. Now, using the same measurements, draw △ $A'B'C'$ so that ∠A' = 30°, △B' = 60°, and AC = 10 cm. Describe how the 10 cm side is positioned.

e. Lastly, again, using the same measurements, draw △ $A''B''C''$ so that ∠A'' = 30°, ∠B'' = 60°, and $B''C''$ = 10 cm. Describe how the 10 cm side is positioned.

f. Are the three drawn triangles identical? Justify your response using measurements.

g. Draw △ $A'''B'''C'''$ so that $\angle B''' = 30°$, $\angle C''' = 60°$, and $B'''C''' = 10$ cm. Is it identical to any of the three triangles already drawn?

h. Draw another triangle that meets the criteria of this challenge. Is it possible to draw any other triangles that would be different from the three drawn above?

EUREKA
MATH

Lesson Summary

The following conditions produce identical triangles:

What Criteria Produce Unique Triangles?

Criteria	Example

Name _____   Date _____

1. A student is given the following three side lengths of a triangle to use to draw a triangle.

    _____

    _____

    _____

    The student uses the longest of the three segments as
    side $\overline{AB}$ of triangle $\triangle ABC$. Explain what the student is
    doing with the two shorter lengths in the work below.
    Then, complete the drawing of the triangle.

    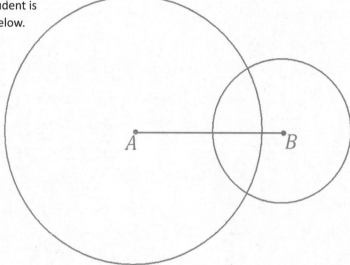

2. Explain why the three triangles constructed in parts (c), (d), and (e) of the Exploratory Challenge were nonidentical.

Necessary Tools

Students need a ruler and protractor to complete the homework assignment.

1. Draw two different equilateral triangles, △ ABC and △ $A'B'C'$. A side length of △ ABC is 2 cm. A side length of △ $A'B'C'$ is 4 cm. Label all sides and angle measurements. Why are your triangles not identical?

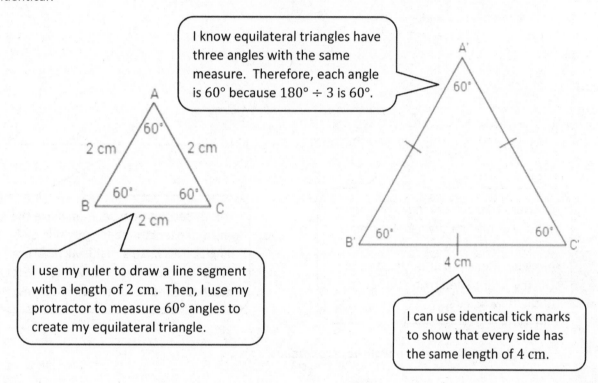

I know equilateral triangles have three angles with the same measure. Therefore, each angle is 60° because 180° ÷ 3 is 60°.

I use my ruler to draw a line segment with a length of 2 cm. Then, I use my protractor to measure 60° angles to create my equilateral triangle.

I can use identical tick marks to show that every side has the same length of 4 cm.

Even though the angles are identical in both triangles, the triangles are not identical because there is no correspondence that matches equal sides to equal sides.

I know that identical triangles must have three identical angles *and* three identical sides.

2. Draw all the isosceles triangles that satisfy the following conditions: one angle measure is 100° and one side has a length of 5 cm. Label all angle and side measurements. How many triangles can be drawn under these conditions?

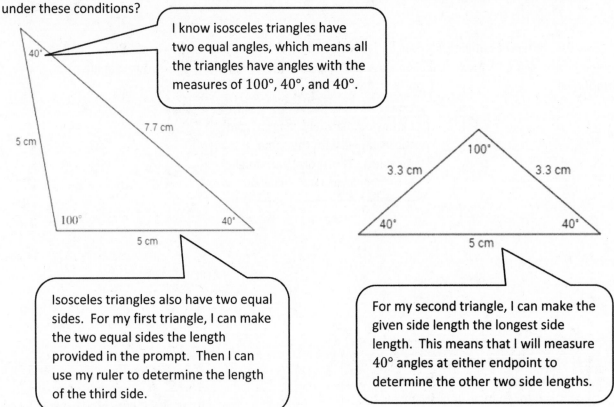

I know isosceles triangles have two equal angles, which means all the triangles have angles with the measures of 100°, 40°, and 40°.

Isosceles triangles also have two equal sides. For my first triangle, I can make the two equal sides the length provided in the prompt. Then I can use my ruler to determine the length of the third side.

For my second triangle, I can make the given side length the longest side length. This means that I will measure 40° angles at either endpoint to determine the other two side lengths.

Only two triangles can be created under these conditions.

Lesson 8:   Drawing Triangles

EUREKA MATH

3. Draw three non-identical triangles so that two angles measure 60° and 80° and one side measures 6 cm. Why are these triangles not identical?

Even though there is a correspondence that will match equal angles to equal angles, these triangles are not identical because there is no correspondence that will match equal sides to equal sides.

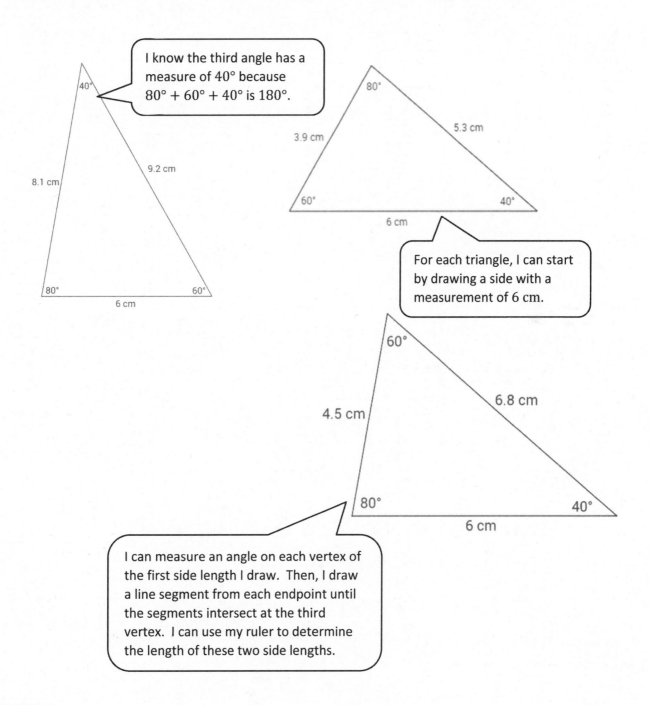

I know the third angle has a measure of 40° because 80° + 60° + 40° is 180°.

For each triangle, I can start by drawing a side with a measurement of 6 cm.

I can measure an angle on each vertex of the first side length I draw. Then, I draw a line segment from each endpoint until the segments intersect at the third vertex. I can use my ruler to determine the length of these two side lengths.

1. Draw three different acute triangles XYZ, $X'Y'Z'$, and $X''Y''Z''$ so that one angle in each triangle is 45°. Label all sides and angle measurements. Why are your triangles not identical?

2. Draw three different equilateral triangles ABC, $A'B'C'$, and $A''B''C''$. A side length of $\triangle ABC$ is 3 cm. A side length of $\triangle A'B'C'$ is 5 cm. A side length of $\triangle A''B''C''$ is 7 cm. Label all sides and angle measurements. Why are your triangles not identical?

3. Draw as many isosceles triangles that satisfy the following conditions: one angle measures 110°, and one side measures 6 cm. Label all angle and side measurements. How many triangles can be drawn under these conditions?

4. Draw three nonidentical triangles so that two angles measure 50° and 60° and one side measures 5 cm.

    a. Why are the triangles not identical?

    b. Based on the diagrams you drew for part (a) and for Problem 2, what can you generalize about the criterion of three given angles in a triangle? Does this criterion determine a unique triangle?

Exploratory Challenge

1. A triangle XYZ exists with side lengths of the segments below. Draw $\triangle X'Y'Z'$ with the same side lengths as $\triangle XYZ$. Use your compass to determine the sides of $\triangle X'Y'Z'$. Use your ruler to measure side lengths. Leave all construction marks as evidence of your work, and label all side and angle measurements.

    Under what condition is $\triangle X'Y'Z'$ drawn? Compare the triangle you drew to two of your peers' triangles. Are the triangles identical? Did the condition determine a unique triangle? Use your construction to explain why. Do the results differ from your predictions?

    X _____ Y

    Y _____ Z

    X _____ Z

2. △ ABC is located below. Copy the sides of the triangle to create △ $A'B'C'$. Use your compass to determine the sides of △ $A'B'C'$. Use your ruler to measure side lengths. Leave all construction marks as evidence of your work, and label all side and angle measurements.

Under what condition is △ $A'B'C'$ drawn? Compare the triangle you drew to two of your peers' triangles. Are the triangles identical? Did the condition determine a unique triangle? Use your construction to explain why.

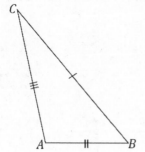

Lesson 9:   Conditions for a Unique Triangle—Three Sides
                          and Two Sides and the Included Angle

EUREKA
MATH

3. A triangle DEF has an angle of 40° adjacent to side lengths of 4 cm and 7 cm. Construct $\triangle\, D'E'F'$ with side lengths $D'E' = 4$ cm, $D'F' = 7$ cm, and included angle $\angle D' = 40°$. Use your compass to draw the sides of $\triangle\, D'E'F'$. Use your ruler to measure side lengths. Leave all construction marks as evidence of your work, and label all side and angle measurements.

    Under what condition is $\triangle\, D'E'F'$ drawn? Compare the triangle you drew to two of your peers' triangles. Did the condition determine a unique triangle? Use your construction to explain why.

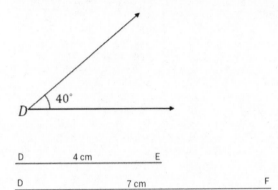

Lesson 9: Conditions for a Unique Triangle—Three Sides
and Two Sides and the Included Angle

4. △ *XYZ* has side lengths *XY* = 2.5 cm, *XZ* = 4 cm, and ∠*X* = 120°. Draw △ *X′Y′Z′* under the same conditions. Use your compass and protractor to draw the sides of △ *X′Y′Z′*. Use your ruler to measure side lengths. Leave all construction marks as evidence of your work, and label all side and angle measurements.

Under what condition is △ *X′Y′Z′* drawn? Compare the triangle you drew to two of your peers' triangles. Are the triangles identical? Did the condition determine a unique triangle? Use your construction to explain why.

Lesson 9:   Conditions for a Unique Triangle—Three Sides
and Two Sides and the Included Angle

Lesson Summary

The following conditions determine a unique triangle:

- Three sides.

- Two sides and an included angle.

Name _____ Date _____

Choose either the three sides condition or the two sides and included angle condition, and explain why the condition determines a unique triangle.

Necessary Tools

Students need a ruler, protractor, and compass to complete this homework assignment.

1. A triangle with side lengths 5 cm, 12 cm, and 13 cm is shown below. Use your compass and ruler to draw a triangle with the same side lengths. Leave all construction marks as evidence of your work, and label all side and angle measurements.

   Under what condition is the triangle drawn? Compare the triangle you drew to the triangle shown below. Are the triangles identical? Did the condition determine a unique triangle? Use your construction to explain why.

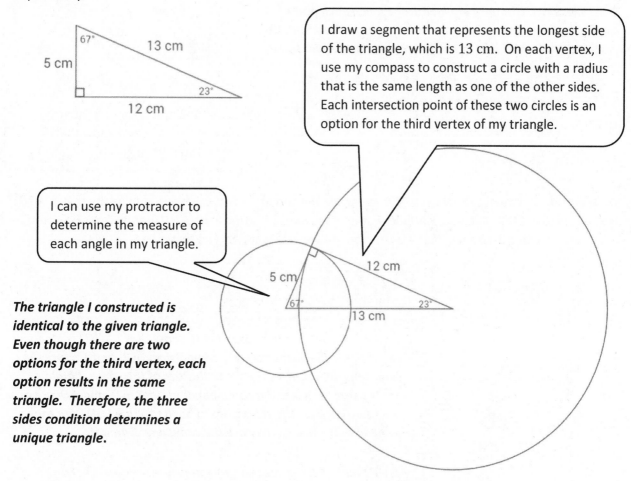

I draw a segment that represents the longest side of the triangle, which is 13 cm. On each vertex, I use my compass to construct a circle with a radius that is the same length as one of the other sides. Each intersection point of these two circles is an option for the third vertex of my triangle.

I can use my protractor to determine the measure of each angle in my triangle.

The triangle I constructed is identical to the given triangle. Even though there are two options for the third vertex, each option results in the same triangle. Therefore, the three sides condition determines a unique triangle.

2. Diagonals $\overline{WY}$ and $\overline{XZ}$ are drawn in square $WXYZ$. Show that $\triangle XYZ$ is identical to $\triangle YZW$, and then use this information to show that the diagonals are equal in length.

I can use the two sides and an included angle condition to show that $\triangle XYZ$ is identical to $\triangle YZW$. I know corresponding sides $\overline{XY}$ and $\overline{YZ}$ are the same length because they are both sides of the same square. For the same reason, corresponding sides $\overline{YZ}$ and $\overline{ZW}$ are also the same length. $\angle XYZ$ and $\angle YZW$ are both right angles, so they have the same measure.

Since these two triangles are identical, I know each pair of corresponding sides must be the same length. Therefore, the diagonals (the third side of each triangle) must be the same length.

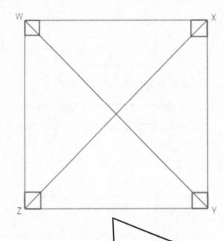

I know that the four sides of a square are all the same length. I also know that all four angles are right angles.

3. Diagonal $\overline{BD}$ is drawn in rhombus $ABCD$. Describe the condition that can be used to justify that $\triangle ABD$ is identical to $\triangle CBD$. Can you conclude that the measures of $\angle ABD$ and $\angle CBD$ are the same? Support your answer with a diagram and an explanation of the correspondence(s) that exists.

I know all four sides of a rhombus have the same length.

I can use the three sides condition to show that $\triangle ABD$ is identical to $\triangle CBD$. I know that corresponding sides $\overline{AB}$ and $\overline{CB}$ are the same length because they are both sides of the same rhombus. For the same reason, the corresponding sides $\overline{AD}$ and $\overline{CD}$ are also the same length. The third sides of each triangle ($\overline{BD}$) are the same length because they are the same line segment.

$\angle ABD$ and $\angle CBD$ have the same measure because they are corresponding angles of two identical triangles.

If two triangles are identical, I know there are corresponding angles that have the same measure.

EUREKA MATH®

1. A triangle with side lengths 3 cm, 4 cm, and 5 cm exists. Use your compass and ruler to draw a triangle with the same side lengths. Leave all construction marks as evidence of your work, and label all side and angle measurements.

    Under what condition is the triangle drawn? Compare the triangle you drew to two of your peers' triangles. Are the triangles identical? Did the condition determine a unique triangle? Use your construction to explain why.

2. Draw triangles under the conditions described below.
    a. A triangle has side lengths 5 cm and 6 cm. Draw two nonidentical triangles that satisfy these conditions. Explain why your triangles are not identical.
    b. A triangle has a side length of 7 cm opposite a 45° angle. Draw two nonidentical triangles that satisfy these conditions. Explain why your triangles are not identical.

3. Diagonal $\overline{BD}$ is drawn in square $ABCD$. Describe what condition(s) can be used to justify that $\triangle ABD$ is identical to $\triangle CBD$. What can you say about the measures of $\angle ABD$ and $\angle CBD$? Support your answers with a diagram and explanation of the correspondence(s) that exists.

4. Diagonals $\overline{BD}$ and $\overline{AC}$ are drawn in square $ABCD$. Show that $\triangle ABC$ is identical to $\triangle BAD$, and then use this information to show that the diagonals are equal in length.

5. Diagonal $\overline{QS}$ is drawn in rhombus $PQRS$. Describe the condition(s) that can be used to justify that $\triangle PQS$ is identical to $\triangle RQS$. Can you conclude that the measures of $\angle PQS$ and $\angle RQS$ are the same? Support your answer with a diagram and explanation of the correspondence(s) that exists.

6. Diagonals $\overline{QS}$ and $\overline{PR}$ are drawn in rhombus $PQRS$ and meet at point T. Describe the condition(s) that can be used to justify that $\triangle PQT$ is identical to $\triangle RQT$. Can you conclude that the line segments PR and QS are perpendicular to each other? Support your answers with a diagram and explanation of the correspondence(s) that exists.

Exploratory Challenge

1. A triangle XYZ has angle measures $\angle X = 30°$ and $\angle Y = 50°$ and included side $XY = 6$ cm. Draw $\triangle X'Y'Z'$ under the same condition as $\triangle XYZ$. Leave all construction marks as evidence of your work, and label all side and angle measurements.

    Under what condition is $\triangle X'Y'Z'$ drawn? Compare the triangle you drew to two of your peers' triangles. Are the triangles identical? Did the condition determine a unique triangle? Use your construction to explain why.

2. A triangle RST has angle measures $\angle S = 90°$ and $\angle T = 45°$ and included side $ST = 7$ cm. Draw $\triangle R'S'T'$ under the same condition. Leave all construction marks as evidence of your work, and label all side and angle measurements. Under what condition is $\triangle R'S'T'$ drawn? Compare the triangle you drew to two of your peers' triangles. Are the triangles identical? Did the condition determine a unique triangle? Use your construction to explain why.

3. A triangle JKL has angle measures $\angle J = 60°$ and $\angle L = 25°$ and side $KL = 5$ cm. Draw $\triangle J'K'L'$ under the same condition. Leave all construction marks as evidence of your work, and label all side and angle measurements. Under what condition is $\triangle J'K'L'$ drawn? Compare the triangle you drew to two of your peers' triangles. Are the triangles identical? Did the condition determine a unique triangle? Use your construction to explain why.

Lesson 10: Conditions for a Unique Triangle—Two Angles and a Given Side

EUREKA MATH®

4. A triangle ABC has angle measures $\angle C = 35°$ and $\angle B = 105°$ and side $AC = 7$ cm. Draw $\triangle A'B'C'$ under the same condition. Leave all construction marks as evidence of your work, and label all side and angle measurements.

   Under what condition is $\triangle A'B'C'$ drawn? Compare the triangle you drew to two of your peers' triangles. Are the triangles identical? Did the condition determine a unique triangle? Use your construction to explain why.

Lesson Summary

The following conditions determine a unique triangle:

- Three sides.
- Two sides and included angle.
- Two angles and the included side.
- Two angles and the side opposite.

EUREKA
MATH®

Name _____   Date _____

1. △ *ABC* has angle measures ∠*A* = 50° and ∠*C* = 90° and side *AB* = 5.5 cm. Draw △ *A'B'C'* under the same condition. Under what condition is △ *A'B'C'* drawn? Use your construction to explain why △ *A'B'C'* is or is not identical to △ *ABC*.

2. △ *PQR* has angle measures ∠*Q* = 25° and ∠*R* = 40° and included side *QR* = 6.5 cm. Draw △ *P'Q'R'* under the same condition. Under what condition is △ *P'Q'R'* drawn? Use your construction to explain why △ *P'Q'R'* is or is not identical to △ *PQR*.

1. In △ ABC, ∠A = 48°, ∠B = 75°, and AB = 4 cm. Draw △ A'B'C' under the same condition as
    △ ABC. Label all side and angle measurements.

    What can you conclude about △ ABC and △ A'B'C'? Justify your response.

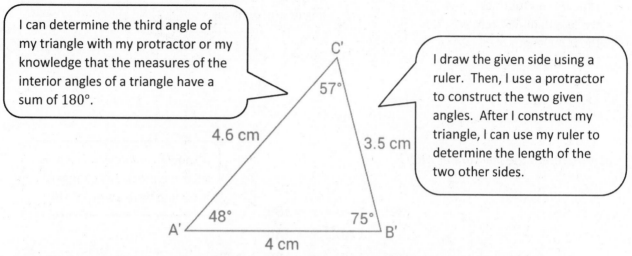

> I can determine the third angle of my triangle with my protractor or my knowledge that the measures of the interior angles of a triangle have a sum of 180°.

> I draw the given side using a ruler. Then, I use a protractor to construct the two given angles. After I construct my triangle, I can use my ruler to determine the length of the two other sides.

Since both triangles are drawn under the same condition, and the two angles and included side condition determines a unique triangle, then both triangles determine the same unique triangle. Therefore, △ ABC and △ A'B'C' are identical.

2. In △ TUV, ∠V = 42°, ∠T = 98°, and UV = 5.5 cm. Draw △ T'U'V' under the same condition as
    △ TUV. Label all side and angle measurements.

    What can you conclude about △ TUV and △ T'U'V'? Justify your response.

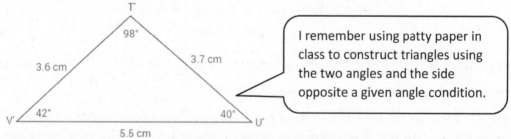

> I remember using patty paper in class to construct triangles using the two angles and the side opposite a given angle condition.

Since both triangles are drawn under the same condition, and the two angles and the side opposite a given angle condition determines a unique triangle, then both triangles determine the same unique triangle. Therefore, △ TUV and △ T'U'V' are identical.

3. In the figure below, points *M*, *N*, and *P* are collinear, and ∠*Q* = ∠*R*. What can be concluded about
    △ *MQN* and △ *PRN*? Justify your response.

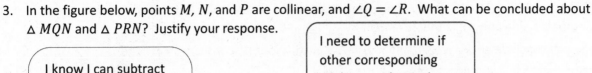
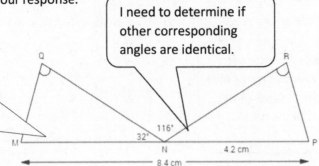

I know I can subtract the length of $\overline{NP}$ from the total length of both segments to determine the length of $\overline{MN}$.

I need to determine if other corresponding angles are identical.

Let x represent the measure of ∠RNP.

I know from previous lessons that the measures of angles on a line have a sum of 180°.

$$32° + 116° + x = 180°$$
$$148° + x = 180°$$
$$148° - 148° + x = 180° - 148°$$
$$x = 32°$$

The measure of ∠RNP is 32°. This means corresponding angles ∠RNP and ∠QNM have the same measure.

Let y represent the measure of $\overline{MN}$.

$$y + 4.2\,\text{cm} = 8.4\,\text{cm}$$
$$y + 4.2\,\text{cm} - 4.2\,\text{cm} = 8.4\,\text{cm} - 4.2\,\text{cm}$$
$$y = 4.2\,\text{cm}$$

The measure of $\overline{MN}$ is 4.2 cm. This means the corresponding sides $\overline{MN}$ and $\overline{PN}$ are the same length.

The diagram indicates ∠Q has the same measure as ∠R.

Therefore, △ MQN and △ PRN are identical because the same measurements in both triangles satisfy the two angles and the side opposite a given angle condition, which means they both determine the same unique triangle.

© 2019 Great Minds®. eureka-math.org

EUREKA
MATH

1. In △ *FGH*, ∠*F* = 42° and ∠*H* = 70°. *FH* = 6 cm. Draw △ *F'G'H'* under the same condition as △ *FGH*. Leave all construction marks as evidence of your work, and label all side and angle measurements.

    What can you conclude about △ *FGH* and △ *F'G'H'*? Justify your response.

2. In △ *WXY*, ∠*Y* = 57° and ∠*W* = 103°. Side *YX* = 6.5 cm. Draw △ *W'X'Y'* under the same condition as △ *WXY*. Leave all construction marks as evidence of your work, and label all side and angle measurements.

    What can you conclude about △ *WXY* and △ *W'X'Y'*? Justify your response.

3. Points *A, Z*, and *E* are collinear, and ∠*B* = ∠*D*. What can be concluded about △ *ABZ* and △ *EDZ*? Justify your answer.

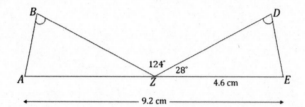

4. Draw △ *ABC* so that ∠*A* has a measurement of 60°, ∠*B* has a measurement of 60°, and $\overline{AB}$ has a length of 8 cm. What are the lengths of the other sides?

5. Draw △ *ABC* so that ∠*A* has a measurement of 30°, ∠*B* has a measurement of 60°, and $\overline{BC}$ has a length of 5 cm. What is the length of the longest side?

Exploratory Challenge 1

   a.   Can any three side lengths form a triangle? Why or why not?

   b.   Draw a triangle according to these instructions:

      ✓   Draw segment AB of length 10 cm in your notebook.

      ✓   Draw segment BC of length 5 cm on one piece of patty paper.

      ✓   Draw segment AC of length 3 cm on the other piece of patty paper.

      ✓   Line up the appropriate endpoint on each piece of patty paper with the matching endpoint on segment AB.

      ✓   Use your pencil point to hold each patty paper in place, and adjust the paper to form $\triangle ABC$.

c.   What do you notice?

d.   What must be true about the sum of the lengths of $\overline{AC}$ and $\overline{BC}$ if the two segments were to just meet? Use your patty paper to verify your answer.

e.   Based on your conclusion for part (d), what if $AC = 3$ cm as you originally had, but $BC = 10$ cm. Could you form $\triangle ABC$?

f.   What must be true about the sum of the lengths of $\overline{AC}$ and $\overline{BC}$ if the two segments were to meet and form a triangle?

Exercise 1

Two sides of $\triangle DEF$ have lengths of 5 cm and 8 cm. What are all the possible whole number lengths for the remaining side?

EUREKA
MATH

Exploratory Challenge 2

a.   Which of the following conditions determine a triangle? Follow the instructions to try to draw △ ABC.
     Segment AB has been drawn for you as a starting point in each case.

   i.   Choose measurements of ∠A and ∠B for △ ABC so that the sum of measurements is greater than 180°.
        Label your diagram.

        Your chosen angle measurements: ∠A =   ∠ B =

        Were you able to form a triangle? Why or why not?

A ———————————————————— B

   ii. Choose measurements of ∠A and ∠B for △ ABC so that the measurement of ∠A is supplementary to the
        measurement of ∠B. Label your diagram.

        Your chosen angle measurements: ∠A =   ∠B =

        Were you able to form a triangle? Why or why not?

A ———————————————————— B

iii. Choose measurements of ∠A and ∠B for △ ABC so that the sum of measurements is less than 180°. Label your diagram.

Your chosen angle measurements: ∠A =   ∠B =

Were you able to form a triangle? Why or why not?

$A\overline{}B$

b. Which condition must be true regarding angle measurements in order to determine a triangle?

c. Measure and label the formed triangle in part (a) with all three side lengths and the angle measurement for ∠C. Now, use a protractor, ruler, and compass to draw △ A′B′C′ with the same angle measurements but side lengths that are half as long.

d. Do the three angle measurements of a triangle determine a unique triangle? Why or why not?

Exercise 2

Which of the following sets of angle measurements determines a triangle?

    a. $30°, 120°$

    b. $125°, 55°$

    c. $105°, 80°$

    d. $90°, 89°$

    e. $91°, 89°$

Choose one example from above that does determine a triangle and one that does not. For each, explain why it does or does not determine a triangle using words and a diagram.

Lesson Summary

- Three lengths determine a triangle provided the largest length is less than the sum of the other two lengths.

- Two angle measurements determine a triangle provided the sum of the two angle measurements is less than 180°.

- Three given angle measurements do not determine a unique triangle.

- Scale drawings of a triangle have equal corresponding angle measurements, but corresponding side lengths are proportional.

Name _____   Date _____

1. What is the maximum and minimum whole number side length for △ XYZ with given side lengths of 3 cm and 5 cm? Please explain why.

2. Jill has not yet studied the angle measurement requirements to form a triangle. She begins to draw side $\overline{AB}$ of △ ABC and considers the following angle measurements for ∠A and ∠B. Describe the drawing that results from each set.

A _____ B

a. 45° and 135°

b. 45° and 45°

c. 45° and 145°

**EUREKA
MATH**®

Lesson 11:   Conditions on Measurements That
                 Determine a Triangle

139

© 2019 Great Minds®. eureka-math.org

Necessary Tools

Students need a ruler and compass to complete the homework assignment.

1. Decide whether each set of three given lengths determines a triangle. For any set of lengths that does determine a triangle, use a ruler and compass to draw the triangle. Label all side lengths. For sets of lengths that do not determine a triangle, write "Does not determine a triangle," and justify your response.

    a. 3 cm, 5 cm, 7 cm

    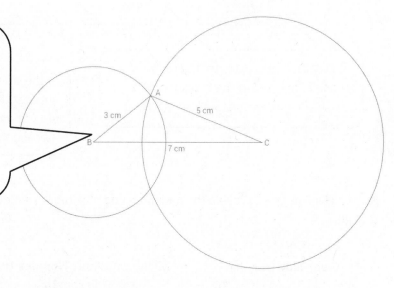

    > I use my ruler to draw the longest side. On one endpoint, I use my compass to draw a circle with a radius of 3 cm. I construct a circle with a radius of 5 cm on the other endpoint. One of the intersection points of these two circles is the third vertex of my triangle.

    b. 6 cm, 13 cm, 5 cm

    These side lengths do not determine a triangle because the two shortest side lengths are too short to create a triangle with a side length of 13 cm.

    > I know that three lengths determine a triangle if the largest length is less than the sum of the other two lengths.
    >
    > $$6 \text{ cm} + 5 \text{ cm} \not> 13 \text{ cm}$$
    >
    > Therefore, these sides lengths will not form a triangle.

2. For each angle measurement below, provide one angle measurement that will determine a triangle and one that will not determine a triangle. Assume that the angles are being drawn to a horizontal segment XY; describe the position of the non-horizontal rays of $\angle X$ and $\angle Y$.

$\angle X$	$\angle Y$: A Measurement That Determines a Triangle	$\angle Y$: A Measurement That Does Not Determine a Triangle
50°	40°	130°
120°	59°	61°

In order for the angles to determine a triangle, the sum of the measures of the two given angles must be less than 180°.

If the sum of the measures of two angles is greater than or equal to 180°, the angles do not determine a triangle because the non-horizontal rays do not intersect.

3. For the given lengths, provide the minimum and maximum whole number side lengths that determine a triangle.

Given Side Lengths	Minimum Whole Number Third Side Length	Maximum Whole Number Third Side Length
4 cm, 5 cm	2 cm	8 cm
2 cm, 14 cm	13 cm	15 cm

I can calculate the minimum possible third length by making the sum of the lengths of the third side and the shorter given side one whole number more than the length of the longest side.

I can calculate the maximum possible third length by calculating the sum of the two given side lengths. I know the length of the third side must be one whole number less than this sum.

Lesson 11: Conditions on Measurements That Determine a Triangle

EUREKA MATH

1. Decide whether each set of three given lengths determines a triangle. For any set of lengths that does determine a triangle, use a ruler and compass to draw the triangle. Label all side lengths. For sets of lengths that do not determine a triangle, write "Does not determine a triangle," and justify your response.

    a. 3 cm, 4 cm, 5 cm

    b. 1 cm, 4 cm, 5 cm

    c. 1 cm, 5 cm, 5 cm

    d. 8 cm, 3 cm, 4 cm

    e. 8 cm, 8 cm, 4 cm

    f. 4 cm, 4 cm, 4 cm

2. For each angle measurement below, provide one angle measurement that will determine a triangle and one that will not determine a triangle. Provide a brief justification for the angle measurements that will not form a triangle. Assume that the angles are being drawn to a horizontal segment AB; describe the position of the non-horizontal rays of angles $\angle A$ and $\angle B$.

$\angle A$	$\angle B$: A Measurement That Determines a Triangle	$\angle B$: A Measurement That *Does Not* Determine a Triangle	Justification for No Triangle
40°			
100°			
90°			
135°			

3. For the given side lengths, provide the minimum and maximum whole number side lengths that determine a triangle.

Given Side Lengths	Minimum Whole Number Third Side Length	Maximum Whole Number Third Side Length
5 cm, 6 cm		
3 cm, 7 cm		
4 cm, 10 cm		
1 cm, 12 cm		

Exploratory Challenge

1. Use your tools to draw △ ABC in the space below, provided $AB = 5$ cm, $BC = 3$ cm, and $\angle A = 30°$. Continue with the rest of the problem as you work on your drawing.

a. What is the relationship between the given parts of △ ABC?

b. Which parts of the triangle can be drawn without difficulty? What makes this drawing challenging?

c. A ruler and compass are instrumental in determining where C is located.
   ✓ Even though the length of segment AC is unknown, extend the ray AC in anticipation of the intersection with segment BC.
   ✓ Draw segment BC with length 3 cm away from the drawing of the triangle.
   ✓ Adjust your compass to the length of $\overline{BC}$.
   ✓ Draw a circle with center B and a radius equal to BC, or 3 cm.

d. How many intersections does the circle make with segment AC? What does each intersection signify?

e. Complete the drawing of △ ABC.

f. Did the results of your drawing differ from your prediction?

Lesson 12:   Unique Triangles—Two Sides and a
              Non-Included Angle

EUREKA
MATH®

2. Now attempt to draw $\triangle DEF$ in the space below, provided $DE = 5$ cm, $EF = 3$ cm, and $\angle F = 90°$. Continue with the rest of the problem as you work on your drawing.

a.   How are these conditions different from those in Exercise 1, and do you think the criteria will determine a unique triangle?

b.   What is the relationship between the given parts of $\triangle DEF$?

c.   Describe how you will determine the position of $\overline{DE}$.

d.   How many intersections does the circle make with $\overline{FD}$?

e.   Complete the drawing of $\triangle DEF$. How is the outcome of $\triangle DEF$ different from that of $\triangle ABC$?

f.   Did your results differ from your prediction?

Lesson 12:   Unique Triangles—Two Sides and a
Non-Included Angle

EUREKA
MATH®

3. Now attempt to draw $\triangle JKL$, provided $KL = 8$ cm, $KJ = 4$ cm, and $\angle J = 120°$. Use what you drew in Exercises 1 and 2 to complete the full drawing.

4. Review the conditions provided for each of the three triangles in the Exploratory Challenge, and discuss the uniqueness of the resulting drawing in each case.

Lesson Summary

Consider a triangle correspondence $\triangle ABC \leftrightarrow \triangle XYZ$ that corresponds to two pairs of equal sides and one pair of equal, non-included angles. If the triangles are not identical, then $\triangle ABC$ can be made to be identical to $\triangle XYZ$ by swinging the appropriate side along the path of a circle with a radius length of that side.

A triangle drawn under the condition of two sides and a non-included angle, where the angle is 90° or greater, creates a unique triangle.

Lesson 12:   Unique Triangles—Two Sides and a
              Non-Included Angle

EUREKA
MATH

Name _____   Date _____

So far, we have learned about four conditions that determine unique triangles: three sides, two sides and an included angle, two angles and an included side, and two angles and the side opposite a given angle.

a. In this lesson, we studied the criterion two sides and a non-included angle. Which case of this criterion determines a unique triangle?

b. Provided $\overline{AB}$ has length 5 cm, $\overline{BC}$ has length 3 cm, and the measurement of $\angle A$ is 30°, draw $\triangle ABC$, and describe why these conditions do not determine a unique triangle.

Necessary Tools

Students need a compass to complete the homework assignment.

1. In the triangle below, two sides and a non-included angle are marked. Use a compass to draw a non-identical triangle that has the same measurements as the marked angle and marked sides. Draw the new triangle on top of the old triangle. What is true about the marked angle in each triangle that results in two non-identical triangles under this condition?

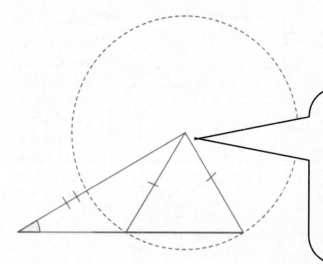

I place the needle of my compass on the top vertex and make the radius the same length as the shortest side of the triangle. The circle intersects with the triangle in another location, which describes where an endpoint of the new triangle is located.

I was able to create a non-identical triangle with the same measurements as the marked angle and marked sides because the non-included angle is an acute angle.

I know that if the non-included angle is smaller than 90°, the two sides and non-included angle condition does not determine a unique triangle.

EUREKA
MATH

Lesson 12: Unique Triangles—Two Sides and a
                Non-Included Angle

153

© 2019 Great Minds®. eureka-math.org

2. A sub-condition of the two sides and non-included angle condition is provided in each row of the following table. Decide whether the information determines a unique triangle. Answer with a *yes, no,* or *maybe* (for a case that may or may not determine a unique triangle).

> I know that the two sides and non-included angle condition always determines a unique triangle when the non-included angle is 90° or larger.

	Condition	Determines a Unique Triangle?
1	Two sides and a non-included 100° angle	Yes
2	Two sides and a non-included 85° angle	Maybe
3	Two sides and a non-included 45° angle, where the side adjacent to the angle is longer than the side opposite the angle	No

> I know that the two sides and non-included angle condition sometimes determines a unique triangle when the non-included angle is acute.

> In order for the two sides and non-included angle condition to determine a unique triangle when the non-included angle is acute, the side adjacent to the angle must be shorter than the side opposite the angle.

EUREKA MATH

1. In each of the triangles below, two sides and a non-included acute angle are marked. Use a compass to draw a nonidentical triangle that has the same measurements as the marked angle and marked sides (look at Exercise 1, part (e) of the Exploratory Challenge as a reference). Draw the new triangle on top of the old triangle. What is true about the marked angles in each triangle that results in two non-identical triangles under this condition?

    a.

    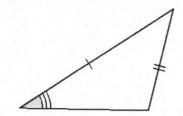

    b.

    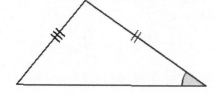

    c.

    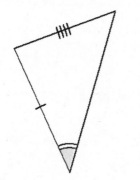

2. Sometimes two sides and a non-included angle of a triangle determine a unique triangle, even if the angle is acute. In the following two triangles, copy the marked information (i.e., two sides and a non-included acute angle), and discover which determines a unique triangle. Measure and label the marked parts.

    In each triangle, how does the length of the marked side adjacent to the marked angle compare with the length of the side opposite the marked angle? Based on your drawings, specifically state when the two sides and acute non-included angle condition determines a unique triangle.

    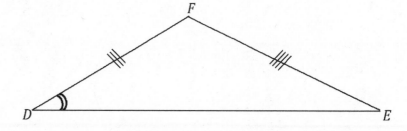

---

3. A sub-condition of the two sides and non-included angle is provided in each row of the following table. Decide whether the information determines a unique triangle. Answer with a *yes, no,* or *maybe* (for a case that may or may not determine a unique triangle).

	Condition	Determines a Unique Triangle?
1	Two sides and a non-included 90° angle.	
2	Two sides and an acute, non-included angle.	
3	Two sides and a non-included 140° angle.	
4	Two sides and a non-included 20° angle, where the side adjacent to the angle is shorter than the side opposite the angle.	
5	Two sides and a non-included angle.	
6	Two sides and a non-included 70° angle, where the side adjacent to the angle is longer than the side opposite the angle.	

4. Choose one condition from the table in Problem 3 that does not determine a unique triangle, and explain why.

5. Choose one condition from the table in Problem 3 that does determine a unique triangle, and explain why.

EUREKA
MATH

Opening Exercise

a.   List all the conditions that determine unique triangles.

b.   How are the terms *identical* and *unique* related?

Each of the following problems gives two triangles. State whether the triangles are *identical, not identical,* or *not necessarily identical*. If the triangles are identical, give the triangle conditions that explain why, and write a triangle correspondence that matches the sides and angles. If the triangles are not identical, explain why. If it is not possible to definitively determine whether the triangles are identical, write "the triangles are not necessarily identical," and explain your reasoning.

Example 1

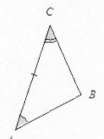
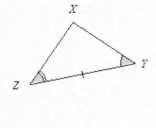

Exercises 1–3

1.

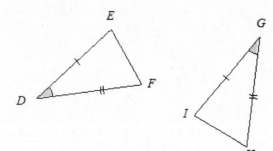

2.

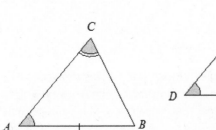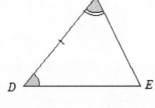

3.

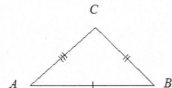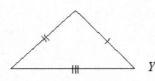

EUREKA
MATH

In Example 2 and Exercises 4–6, three pieces of information are given for $\triangle ABC$ and $\triangle XYZ$. Draw, freehand, the two triangles (do not worry about scale), and mark the given information. If the triangles are identical, give a triangle correspondence that matches equal angles and equal sides. Explain your reasoning.

Example 2

$AB = XZ$, $AC = XY$, $\angle A = \angle X$

Exercises 4–6

4. $\angle A = \angle Z$, $\angle B = \angle Y$, $AB = YZ$

5. $\angle A = \angle Z, \angle B = \angle Y, BC = XY$

6. $\angle A = \angle Z, \angle B = \angle Y, BC = XZ$

© 2019 Great Minds®. eureka-math.org

EUREKA
MATH®

Lesson Summary

The measurement and arrangement (and correspondence) of the parts in each triangle play a role in determining whether two triangles are identical.

Name _____ Date _____

$\angle A$ and $\angle D$ are equal in measure. Draw two triangles around each angle, and mark parts appropriately so that the triangles are identical; use $\angle A$ and $\angle D$ as part of the chosen condition. Write a correspondence for the triangles.

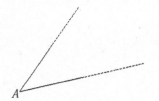

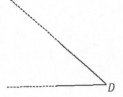

In each of the following problems, two triangles are given. State whether the triangles are *identical, not identical*, or *not necessarily identical*. If the triangles are identical, give the triangle conditions that explain why, and write a triangle correspondence that matches the sides and angles. If the triangles are not identical, explain why. If it is not possible to definitively determine whether the triangles are identical, write "the triangles are not necessarily identical," and explain your reasoning.

1

> I see two sets of identical angles and one set of identical sides. The location of the angles and sides follow the two angles and included side condition.

> I know that ∠A corresponds to ∠Z, and ∠B corresponds to ∠Y. I also know that side $\overline{AB}$ corresponds to $\overline{ZY}$. I can use this information to write the correspondence.

The triangles are identical by the two angles and included side condition. The correspondence △ABC ↔ △ZYX matches two equal pairs of angles and one equal pair of included sides. Since both triangles have parts under the condition of the same measurement, the triangles must be identical.

2.

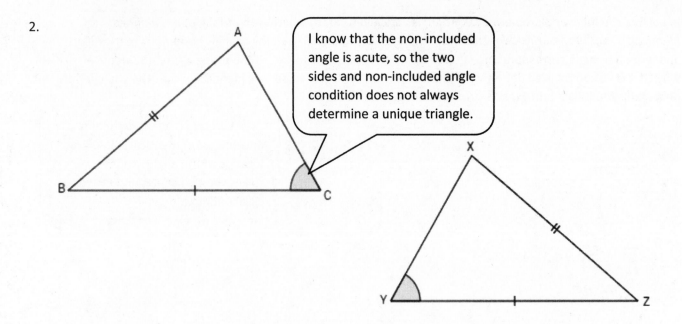

I know that the non-included angle is acute, so the two sides and non-included angle condition does not always determine a unique triangle.

The triangles are not necessarily identical by the two sides and a non-included angle condition. I would need more information about the given sides in order to determine whether or not the two triangles are identical.

I know that the side adjacent to the given angle must be shorter than the side opposite the given angle in order for these triangles to be identical. However, I cannot determine which side is longer with the information given.

In the following problems, three pieces of information are given for △ *ABC* and △ *XYZ*. Draw, freehand, the two triangles (do not worry about scale), and mark the given information. If the triangles are identical, give a triangle correspondence that matches equal angles and equal sides. Explain your reasoning.

Lesson 13:   Checking for Identical Triangles

EUREKA MATH

3. ∠A = ∠Z, ∠B = ∠X, and AB = ZX

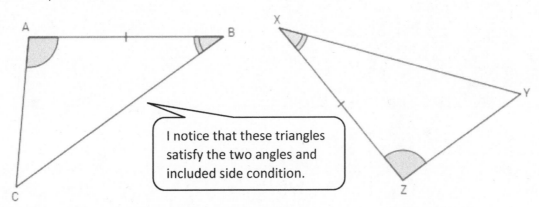

I notice that these triangles satisfy the two angles and included side condition.

These triangles are identical by the two angles and included side condition. The triangle correspondence △ABC ↔ △ZXY matches the two pairs of equal angles and one pair of equal sides. Since both triangles have parts under the condition of the same measurement, the triangles must be identical.

When I write the correspondence, I need to make sure the equal angles and equal sides are in the same location for each triangle.

4. AB = XY, BC = XZ, and ∠C = ∠Y

Originally I thought these triangles satisfied the two sides and non-included angle condition. However, I notice that the side adjacent to ∠C is not equal to the side that is adjacent to ∠Y.

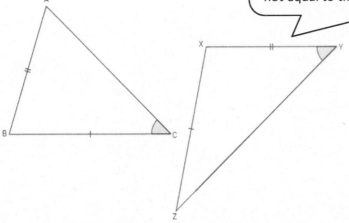

These triangles are not necessarily identical. In △ABC, the marked angle is adjacent to the side marked with one tick mark. In △XYZ, the marked angle is adjacent to the side marked with two tick marks. Since the sides adjacent to the equal angles are not equal in length, the triangles do not fit any of the conditions that determine a unique triangle.

In each of the following four problems, two triangles are given. State whether the triangles are *identical, not identical,* or *not necessarily identical.* If the triangles are identical, give the triangle conditions that explain why, and write a triangle correspondence that matches the sides and angles. If the triangles are not identical, explain why. If it is not possible to definitively determine whether the triangles are identical, write "the triangles are not necessarily identical," and explain your reasoning.

1.

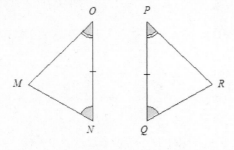

2.

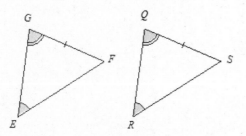

3.

4.

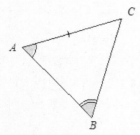
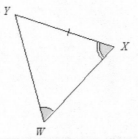

For Problems 5–8, three pieces of information are given for $\triangle ABC$ and $\triangle YZX$. Draw, freehand, the two triangles (do not worry about scale), and mark the given information. If the triangles are identical, give a triangle correspondence that matches equal angles and equal sides. Explain your reasoning.

5. $AB = YZ,\ BC = ZX,\ AC = YX$

6. $AB = YZ,\ BC = ZX,\ \angle C = \angle Y$

7. $AB = XZ,\ \angle A = \angle Z,\ \angle C = \angle Y$

8. $AB = XY,\ AC = YZ,\ \angle C = \angle Z$ (Note that both angles are obtuse.)

EUREKA
MATH

In each of the following problems, determine whether the triangles are *identical, not identical*, or *not necessarily identical*; justify your reasoning. If the relationship between the two triangles yields information that establishes a condition, describe the information. If the triangles are identical, write a triangle correspondence that matches the sides and angles.

Example 1

What is the relationship between the two triangles below?

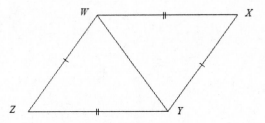

Exercises 1–2

1. Are the triangles identical? Justify your reasoning.

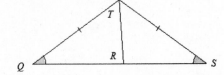

2. Are the triangles identical? Justify your reasoning.

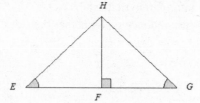

Example 2

Are the triangles identical? Justify your reasoning.

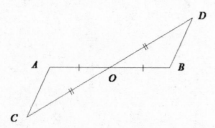

EUREKA
MATH

Exercises 3–4

3. Are the triangles identical? Justify your reasoning.

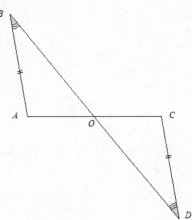

4. Are the triangles identical? Justify your reasoning.

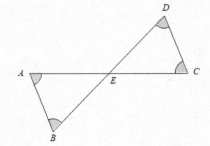

EUREKA
MATH

© 2019 Great Minds®. eureka-math.org

Exercises 5–8

5. Are the triangles identical? Justify your reasoning.

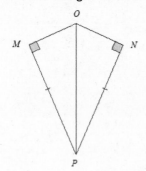

6. Are the triangles identical? Justify your reasoning.

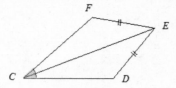

Lesson 14:   Checking for Identical Triangles

EUREKA MATH

7. Are the triangles identical? Justify your reasoning.

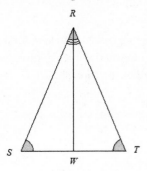

8. Create your own labeled diagram and set of criteria for a pair of triangles. Ask a neighbor to determine whether the triangles are identical based on the provided information.

Lesson Summary

In deciding whether two triangles are identical, examine the structure of the diagram of the two triangles to look for a relationship that might reveal information about corresponding parts of the triangles. This information may determine whether the parts of the triangle satisfy a particular condition, which might determine whether the triangles are identical.

Name _____ Date _____

Are △ *DEF* and △ *DGF* identical, not identical, or not necessarily identical? Justify your reasoning. If the relationship between the two triangles yields information that establishes a condition, describe the information. If the triangles are identical, write a triangle correspondence that matches the sides and angles.

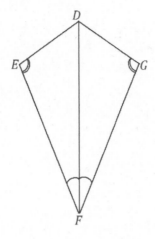

In the following problems, determine whether the triangles are *identical, not identical, or not necessarily identical*; justify your reasoning. If the relationship between the two triangles yields information that establishes a condition, describe the information. If the triangles are identical, write a triangle correspondence that matches the sides and angles.

1.

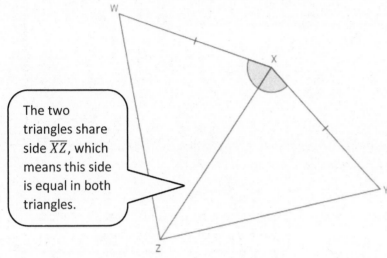

The two triangles share side $\overline{XZ}$, which means this side is equal in both triangles.

These triangles are identical by the two sides and the included angle condition. The triangle correspondence △ WXZ ↔ △ YXZ matches two pairs of equal sides and one pair of equal angles.

One of the equal pairs of sides is shared side $\overline{XZ}$.

2.

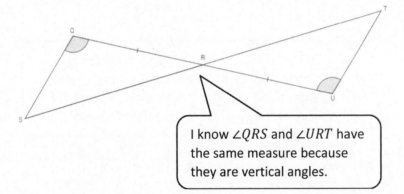

I know ∠QRS and ∠URT have the same measure because they are vertical angles.

The two triangles are identical by the two angles and the included side condition. The triangle correspondence △ QRS ↔ △ URT matches the two pairs of equal angles and one pair of equal sides.

One of the pairs of equal angles are vertical angles.

3.

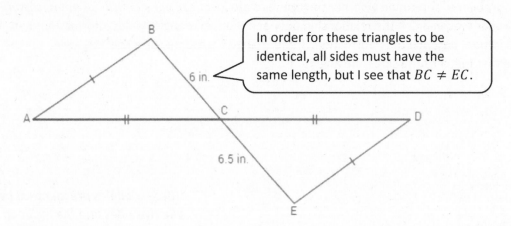

In order for these triangles to be identical, all sides must have the same length, but I see that $BC \neq EC$.

The two triangles are not identical because the correspondence that matches the two marked equal pairs of sides also matches sides $\overline{BC}$ and $\overline{EC}$, which are not equal in length.

The triangles are not identical, so I cannot write a triangle correspondence.

4.

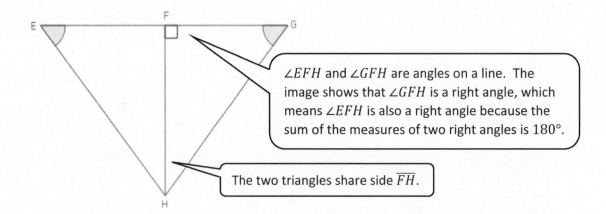

$\angle EFH$ and $\angle GFH$ are angles on a line. The image shows that $\angle GFH$ is a right angle, which means $\angle EFH$ is also a right angle because the sum of the measures of two right angles is 180°.

The two triangles share side $\overline{FH}$.

These two triangles are identical by the two angles and side opposite a given angle condition. The correspondence △EFH ↔ △GFH matches the two pairs of equal angles and the one pair of equal sides.

One of the pairs of equal angles is $\angle EFH$ and $\angle GFH$ because they are both right angles.

EUREKA MATH®

In the following problems, determine whether the triangles are *identical, not identical,* or *not necessarily identical*; justify your reasoning. If the relationship between the two triangles yields information that establishes a condition, describe the information. If the triangles are identical, write a triangle correspondence that matches the sides and angles.

1.

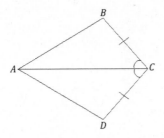

2.

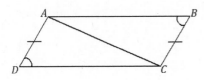

3.

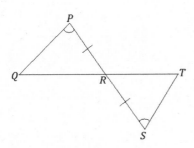

4.

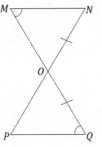

5.

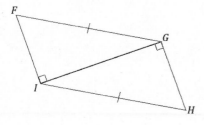

6.

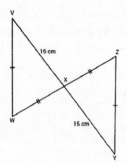

7.

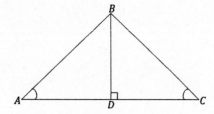

8.   Are there any identical triangles in this diagram?

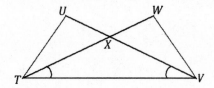

9.

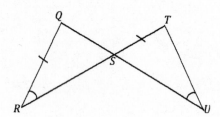

10.

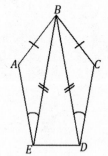

EUREKA MATH

Example 1

A triangular fence with two equal angles, $\angle S = \angle T$, is used to enclose some sheep. A fence is constructed inside the triangle that exactly cuts the other angle into two equal angles: $\angle SRW = \angle TRW$. Show that the gates, represented by $\overline{SW}$ and $\overline{WT}$, are the same width.

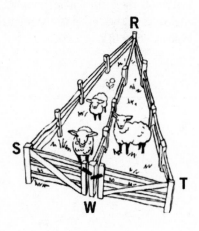

Example 2

In $\triangle ABC$, $AC = BC$, and $\triangle ABC \leftrightarrow A\,B'A'C'$. John says that the triangle correspondence matches two sides and the included angle and shows that $\angle A = \angle B'$. Is John correct?

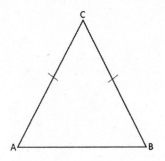
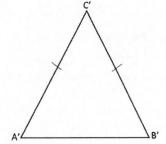

Exercises 1–4

1. Mary puts the center of her compass at the vertex O of the angle and locates points A and B on the sides of the angle. Next, she centers her compass at each of A and B to locate point C. Finally, she constructs the ray $\overrightarrow{OC}$. Explain why $\angle BOC = \angle AOC$.

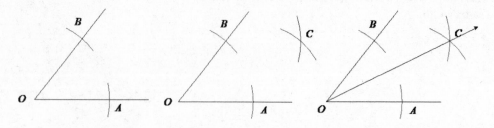

2. Quadrilateral $ACBD$ is a model of a kite. The diagonals $\overline{AB}$ and $\overline{CD}$ represent the sticks that help keep the kite rigid.

   a. John says that $\angle ACD = \angle BCD$. Can you use identical triangles to show that John is correct?

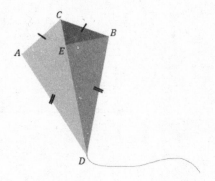

EUREKA
MATH

b. Jill says that the two sticks are perpendicular to each other. Use the fact that $\angle ACD = \angle BCD$ and what you know about identical triangles to show $\angle AEC = 90°$.

c. John says that Jill's triangle correspondence that shows the sticks are perpendicular to each other also shows that the sticks cross at the midpoint of the horizontal stick. Is John correct? Explain.

3. In $\triangle ABC$, $\angle A = \angle B$, and $\triangle ABC \leftrightarrow \triangle B'A'C'$. Jill says that the triangle correspondence matches two angles and the included side and shows that $AC = B'C'$. Is Jill correct?

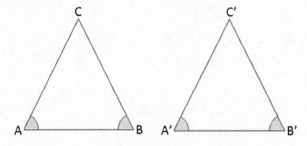

4. Right triangular corner flags are used to mark a soccer field. The
    vinyl flags have a base of 40 cm and a height of 14 cm.

    a. Mary says that the two flags can be obtained by cutting a
        rectangle that is 40 cm × 14 cm on the diagonal. Will that
        create two identical flags? Explain.

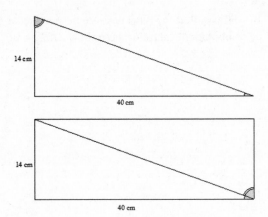

    b. Will measures the two non-right angles on a flag and adds the measurements together. Can you explain,
        without measuring the angles, why his answer is 90°?

Lesson Summary

- In deciding whether two triangles are identical, examine the structure of the diagram of the two triangles to look for a relationship that might reveal information about corresponding parts of the triangles. This information may determine whether the parts of the triangle satisfy a particular condition, which might determine whether the triangles are identical.

- Be sure to identify and label all known measurements, and then determine if any other measurements can be established based on knowledge of geometric relationships.

Name _____   Date _____

Alice is cutting wrapping paper to size to fit a package. How should she cut the rectangular paper into two triangles to ensure that each piece of wrapping paper is the same? Use your knowledge of conditions that determine unique triangles to justify that the pieces resulting from the cut are the same.

1. Ms. Thompson wants to cut different sheets of paper into four equal triangles for a class activity. She first cuts the paper into equal halves in the shape of rectangles, and then she cuts each rectangle along a diagonal.

    Did Ms. Thompson cut the paper into 4 equal pieces? Explain.

    > Each of the four triangles have two sides and one right angle from the rectangles from the first cut. Due to properties of rectangles, I know there is a correspondence between all four triangles that matches two pairs of equal sides and one pair of equal angles.

    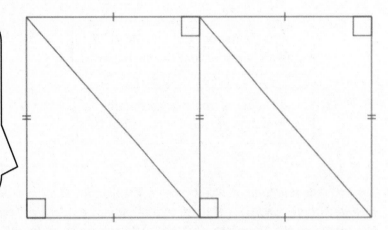

Ms. Thompson did cut the piece of paper into four identical triangles. The first cut Ms. Thompson made resulted in two equal rectangles, which means the corresponding sides have the same length. I also know that all four angles in a rectangle are right angles. Therefore, the four triangles are identical due to the two sides and an included angle condition.

2. The bridge below, which crosses a road, is built out of two triangular supports. The point W lies on $\overline{VX}$. The beams represented by $\overline{YW}$ and $\overline{ZW}$ are equal in length, and the beams represented by $\overline{YV}$ and $\overline{ZX}$ are equal in length. If the supports were constructed so that $\angle Y$ and $\angle Z$ are equal in measurement, is point W the midpoint of $\overline{VX}$? Explain.

If W is the midpoint of $\overline{VX}$, then $\overline{VW}$ and $\overline{XW}$ must have the same length.

I add marks on the image from the information provided in the prompt.

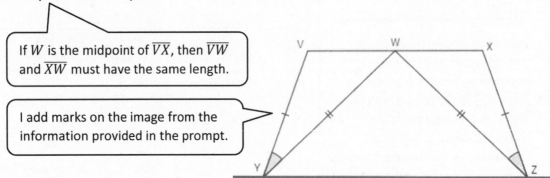

Yes, W is the midpoint of $\overline{VX}$. I know $\triangle WVY$ and $\triangle WXZ$ are identical triangles due to the two sides and included angle condition. If these two triangles are identical, that means the corresponding sides have the same length. Therefore, $\overline{VW}$ and $\overline{XW}$ are the same length.

EUREKA
MATH

1. Jack is asked to cut a cake into 8 equal pieces. He first cuts it into equal fourths in the shape of rectangles, and then he cuts each rectangle along a diagonal.

    Did he cut the cake into 8 equal pieces? Explain.

    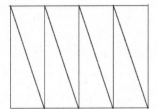

2. The bridge below, which crosses a river, is built out of two triangular supports. The point M lies on $\overline{BC}$. The beams represented by $\overline{AM}$ and $\overline{DM}$ are equal in length, and the beams represented by $\overline{AB}$ and $\overline{DC}$ are equal in length. If the supports were constructed so that $\angle A$ and $\angle D$ are equal in measurement, is point M the midpoint of $\overline{BC}$? Explain.

    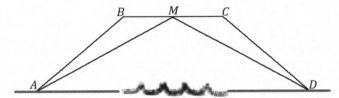

Example 1

Consider a ball B. Figure 3 shows one possible slice of B.

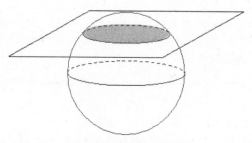

Figure 3. A Slice of Ball B

a. What figure does the slicing plane form? Students may choose their method of representation of the slice (e.g., drawing a 2D sketch, a 3D sketch, or describing the slice in words).

b. Will all slices that pass through B be the same size? Explain your reasoning.

c. How will the plane have to meet the ball so that the plane section consists of just one point?

Example 2

The right rectangular prism in Figure 4 has been sliced with a plane parallel to face $ABCD$. The resulting slice is a rectangular region that is identical to the parallel face.

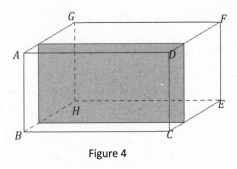

Figure 4

a. Label the vertices of the rectangular region defined by the slice as $WXYZ$.

b. To which other face is the slice parallel and identical?

c. Based on what you know about right rectangular prisms, which faces must the slice be perpendicular to?

Exercise 1

Discuss the following questions with your group.

1. The right rectangular prism in Figure 5 has been sliced with a plane parallel to face $LMON$.

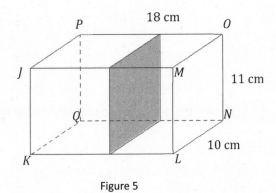

Figure 5

   a. Label the vertices of the rectangle defined by the slice as $RSTU$.

   b. What are the dimensions of the slice?

   c. Based on what you know about right rectangular prisms, which faces must the slice be perpendicular to?

EUREKA MATH

Example 3

The right rectangular prism in Figure 6 has been sliced with a plane perpendicular to $BCEH$. The resulting slice is a rectangular region with a height equal to the height of the prism.

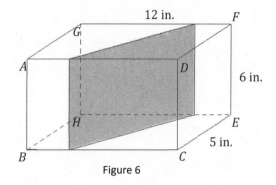
Figure 6

a. Label the vertices of the rectangle defined by the slice as $WXYZ$.

b. To which other face is the slice perpendicular?

c. What is the height of rectangle $WXYZ$?

d. Joey looks at $WXYZ$ and thinks that the slice may be a parallelogram that is not a rectangle. Based on what is known about how the slice is made, can he be right? Justify your reasoning.

Exercises 2–6

In the following exercises, the points at which a slicing plane meets the edges of the right rectangular prism have been marked. Each slice is either parallel or perpendicular to a face of the prism. Use a straightedge to join the points to outline the rectangular region defined by the slice, and shade in the rectangular slice.

2. A slice parallel to a face

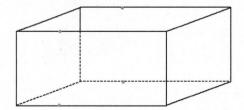

3.   A slice perpendicular to a face

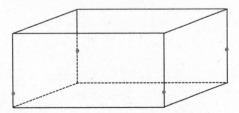

4.   A slice perpendicular to a face

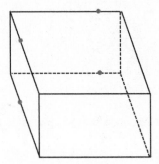

In Exercises 5–6, the dimensions of the prisms have been provided. Use the dimensions to sketch the slice from each prism, and provide the dimensions of each slice.

5.   A slice parallel to a face

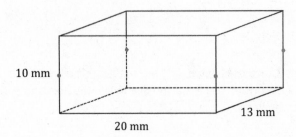

10 mm

20 mm

13 mm

EUREKA
MATH

6.   A slice perpendicular to a face

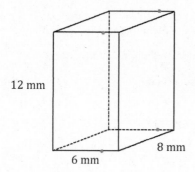

12 mm

6 mm

8 mm

Lesson Summary

- A slice, also known as a plane section, consists of all the points where the plane meets the figure.

- A slice made parallel to a face in a right rectangular prism will be parallel and identical to the face.

- A slice made perpendicular to a face in a right rectangular prism will be a rectangular region with a height equal to the height of the prism.

Name _____ Date _____

In the following figures, use a straightedge to join the points where a slicing plane meets with a right rectangular prism to outline the slice.

    i.   Label the vertices of the rectangular slice $WXYZ$.

    ii.   State any known dimensions of the slice.

    iii. Describe two relationships slice $WXYZ$ has in relation to faces of the right rectangular prism.

1.

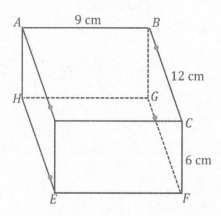

2.

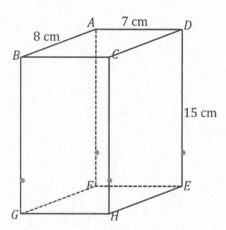

A right rectangular prism is shown along with line segments that lie in a face. For segments a and b, draw and give the approximate dimensions of the slice that results when the slicing plane contains the given line segment and is perpendicular to the face that contains the line segment.

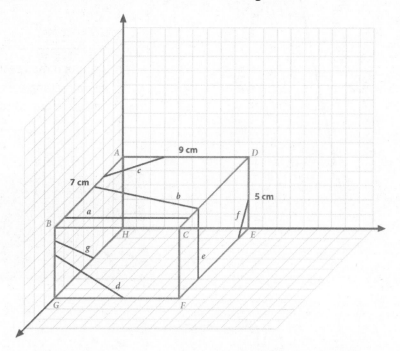

a.

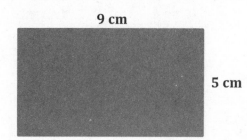

I see that the line segment a is perpendicular to $DCFE$ and $ABGH$. This means that a rectangular slice will be created. I know the dimensions will match the dimensions of different lengths of the prism, in this case 9 cm by 5 cm.

b.

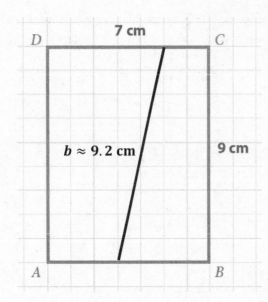

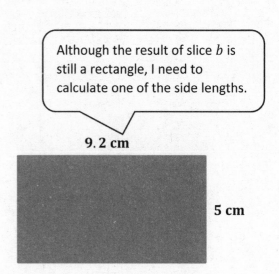

Although the result of slice b is still a rectangle, I need to calculate one of the side lengths.

9.2 cm

5 cm

Lesson 16:   Slicing a Right Rectangular Prism with a Plane

EUREKA MATH

A right rectangular prism is shown along with line segments that lie in a face. For each line segment, draw and give the approximate dimensions of the slice that results when the slicing plane contains the given line segment and is perpendicular to the face that contains the line segment.

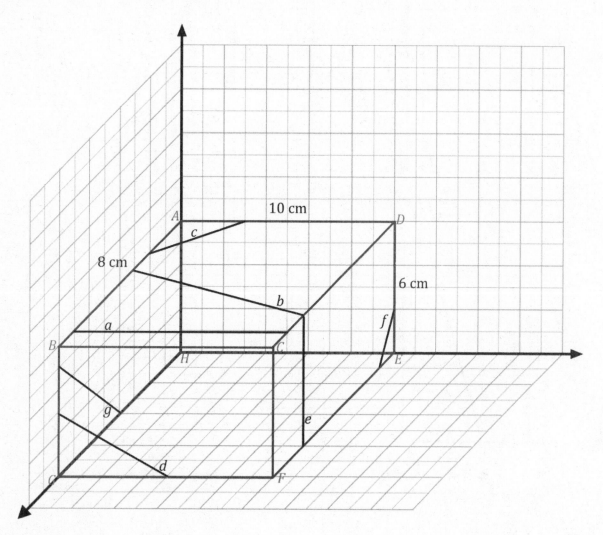

a.

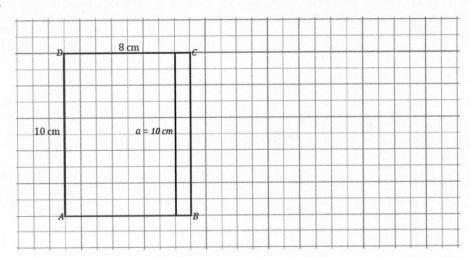

b.

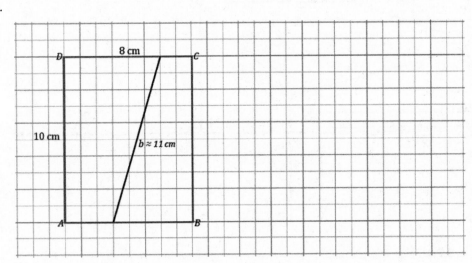

c.

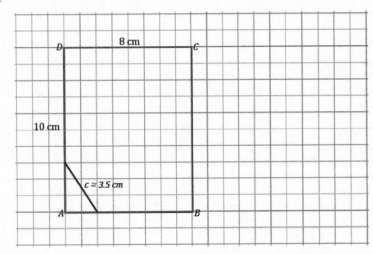

Lesson 16:   Slicing a Right Rectangular Prism with a Plane

EUREKA
MATH®

d.

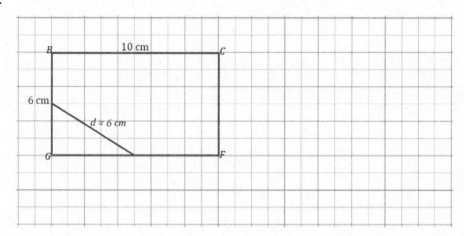

e.

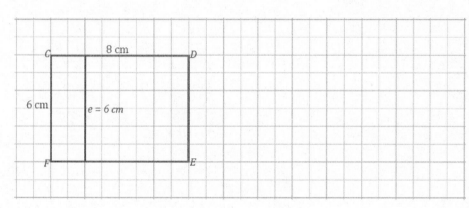

f.

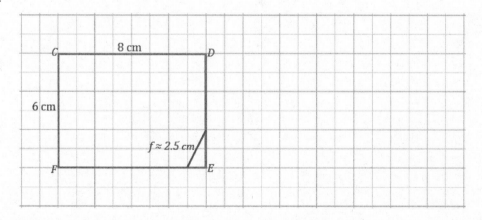

g.

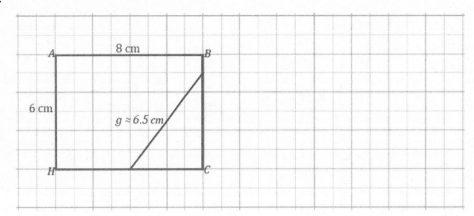

EUREKA
MATH®

Opening

RECTANGULAR PYRAMID: Given a rectangular region B in a plane E, and a point V not in E, the *rectangular pyramid with base B and vertex V* is the collection of all segments VP for any point P in B. It can be shown that the planar region defined by a side of the base B and the vertex V is a triangular region called a *lateral face*.

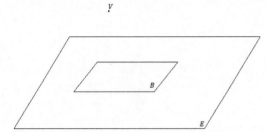

A rectangular region B in a plane E and a point V not in E

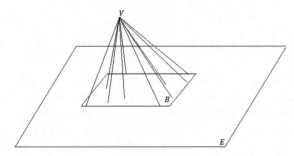

The rectangular pyramid is determined by the collection of all segments VP for any point P in B; here $\overline{VP}$ is shown for a total of 10 points.

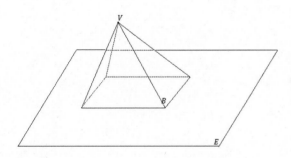

The rectangular pyramid is a solid once the collection of all segments VP for any point P in B are taken. The pyramid has a total of five faces: four lateral faces and a base.

If the vertex lies on the line perpendicular to the base at its center (i.e., the intersection of the rectangle's diagonals), the pyramid is called a *right rectangular pyramid*. The example of the rectangular pyramid above is not a right rectangular pyramid, as evidenced in this image. The perpendicular from V does not meet at the intersection of the diagonals of the rectangular base B.

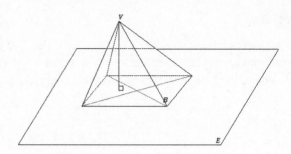

The following is an example of a right rectangular pyramid. The opposite lateral faces are identical isosceles triangles.

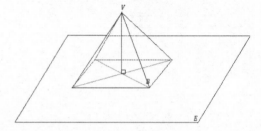

Example 1

Use the models you built to assist in a sketch of a pyramid. Though you are sketching from a model that is opaque, use dotted lines to represent the edges that cannot be seen from your perspective.

Lesson 17:   Slicing a Right Rectangular Pyramid with a Plane

EUREKA MATH®

Example 2

Sketch a right rectangular pyramid from three vantage points: (1) from directly over the vertex, (2) from facing straight on to a lateral face, and (3) from the bottom of the pyramid. Explain how each drawing shows each view of the pyramid.

Example 3

Assume the following figure is a top-down view of a rectangular pyramid. Make a reasonable sketch of any two adjacent lateral faces. What measurements must be the same between the two lateral faces? Mark the equal measurement. Justify your reasoning for your choice of equal measurements.

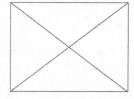

Example 4

a. A slicing plane passes through segment a parallel to base B of the right rectangular pyramid below. Sketch what the slice will look like into the figure. Then sketch the resulting slice as a two-dimensional figure. Students may choose how to represent the slice (e.g., drawing a 2D or 3D sketch or describing the slice in words).

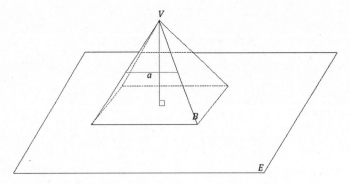

b. What shape does the slice make? What is the relationship between the slice and the rectangular base of the pyramid?

EUREKA
MATH

Example 5

A slice is to be made along segment a perpendicular to base B of the right rectangular pyramid below.

a. Which of the following figures shows the correct slice? Justify why each of the following figures is or is not a correct diagram of the slice.

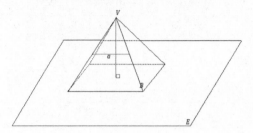

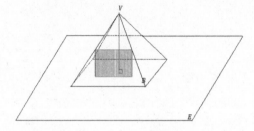

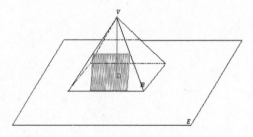

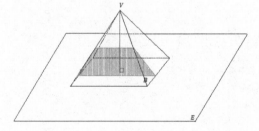

b.   A slice is taken through the vertex of the pyramid perpendicular to the base. Sketch what the slice will look like into the figure. Then, sketch the resulting slice itself as a two-dimensional figure.

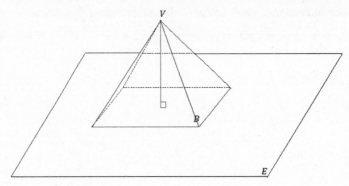

Lesson Summary

- A rectangular pyramid differs from a right rectangular pyramid because the vertex of a right rectangular pyramid lies on the line perpendicular to the base at its center whereas a pyramid that is not a right rectangular pyramid will have a vertex that is not on the line perpendicular to the base at its center.

- Slices made parallel to the base of a right rectangular pyramid are scale drawings of the rectangular base of the pyramid.

Name _____ Date _____

Two copies of the same right rectangular pyramid are shown below. Draw in the slice along segment c perpendicular to the base and the slice along segment c parallel to the base. Then, sketch the resulting slices as two-dimensional figures.

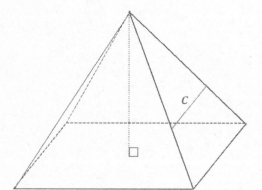   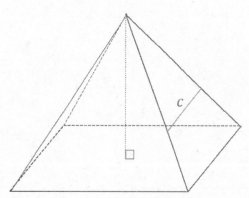

Slice Perpendicular to Base   Slice Parallel to Base

A side view of a right rectangular pyramid is given. The line segments lie in the lateral faces.

a. For segment *a*, sketch the resulting slice from slicing the right rectangular pyramid with a slicing plane that contains the segment and is parallel to the base.

b. For segments *b* and *c*, sketch the resulting slice from slicing the right rectangular pyramid with a slicing plane that contains the line segment and is perpendicular to the base.

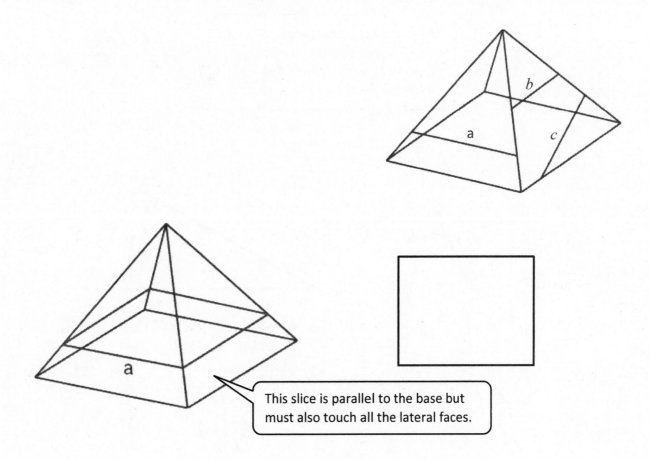

This slice is parallel to the base but must also touch all the lateral faces.

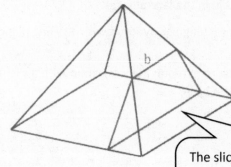

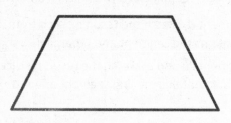

The slice must touch the lateral faces and be perpendicular to the base. I can also attempt to draw the slice within the prism for a challenge.

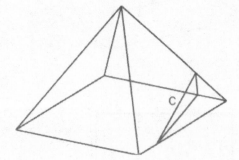

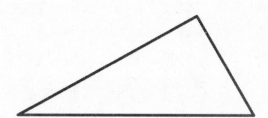

A side view of a right rectangular pyramid is given. The line segments lie in the lateral faces.

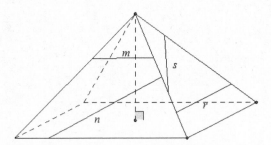

a.   For segments n, s, and r, sketch the resulting slice from slicing the right rectangular pyramid with a slicing plane that contains the line segment and is perpendicular to the base.

b.   For segment m, sketch the resulting slice from slicing the right rectangular pyramid with a slicing plane that contains the segment and is parallel to the base.

Note: To challenge yourself, you can try drawing the slice into the pyramid.

c.   A top view of a right rectangular pyramid is given. The line segments lie in the base face. For each line segment, sketch the slice that results from slicing the right rectangular pyramid with a plane that contains the line segment and is perpendicular to the base.

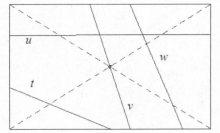

With your group, discuss whether a right rectangular prism can be sliced at an angle so that the resulting slice looks like the figure in Figure 1. If it is possible, draw an example of such a slice into the following prism.

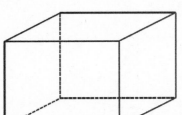

Figure 1

Exercise 1

a. With your group, discuss how to slice a right rectangular prism so that the resulting slice looks like the figure in Figure 2. Justify your reasoning.

Figure 2

b. With your group, discuss how to slice a right rectangular prism so that the resulting slice looks like the figure in Figure 3. Justify your reasoning.

Figure 3

Example 2

With your group, discuss whether a right rectangular prism can be sliced at an angle so that the resulting slice looks like the figure in Figure 4. If it is possible, draw an example of such a slice into the following prism.

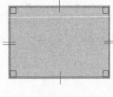

Figure 4

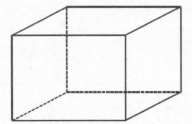

Exercise 2

In Example 2, we discovered how to slice a right rectangular prism to makes the shapes of a rectangle and a parallelogram. Are there other ways to slice a right rectangular prism that result in other quadrilateral-shaped slices?

Example 3

a. If slicing a plane through a right rectangular prism so that the slice meets the three faces of the prism, the resulting slice is in the shape of a triangle; if the slice meets four faces, the resulting slice is in the shape of a quadrilateral. Is it possible to slice the prism in a way that the region formed is a pentagon (as in Figure 5)? A hexagon (as in Figure 6)? An octagon (as in Figure 7)?

Figure 5

Figure 6

Figure 7

Lesson 18:   Slicing on an Angle

EUREKA
MATH

b. Draw an example of a slice in a pentagon shape and a slice in a hexagon shape.

Example 4

a. With your group, discuss whether a right rectangular pyramid can be sliced at an angle so that the resulting slice looks like the figure in Figure 8. If it is possible, draw an example of such a slice into the following pyramid.

Figure 8

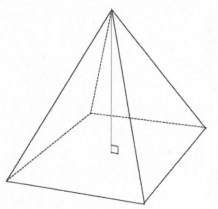

b. With your group, discuss whether a right rectangular pyramid can be sliced at an angle so that the resulting slice looks like the figure in Figure 9. If it is possible, draw an example of such a slice into the pyramid above.

Figure 9

Lesson Summary

- Slices made at an angle are neither parallel nor perpendicular to a base.

- There cannot be more sides to the polygonal region of a slice than there are faces of the solid.

Lesson 18:   Slicing on an Angle

EUREKA
MATH

Name _____   Date _____

Draw a slice that has the maximum possible number of sides for each solid. Explain how you got your answer.

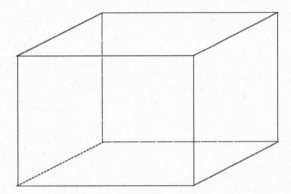

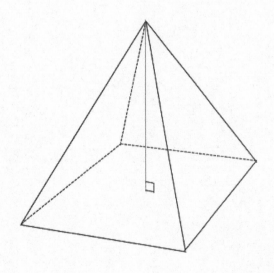

1. Draw a slice into a right rectangular prism at an angle in the form of the provided shape, and draw each slice as a 2D shape.

    a. A quadrilateral

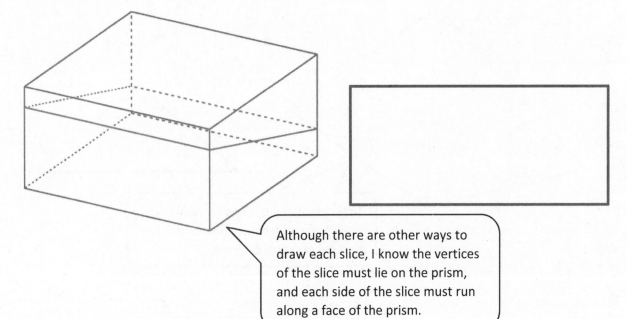

Although there are other ways to draw each slice, I know the vertices of the slice must lie on the prism, and each side of the slice must run along a face of the prism.

    b. A hexagon

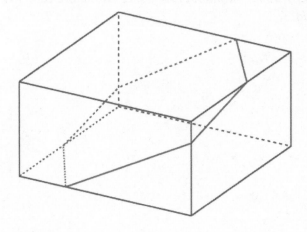
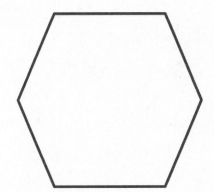

2. Draw a slice on an angle into the right rectangular pyramid below in the form of a triangle, then draw the slice as a 2D shape.

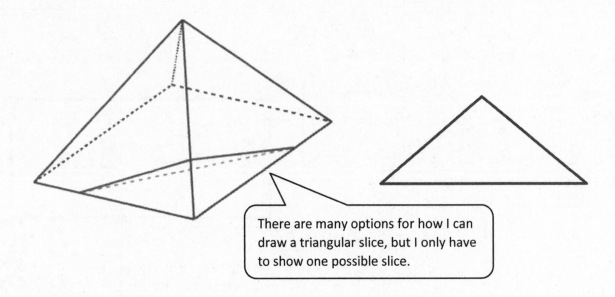

There are many options for how I can draw a triangular slice, but I only have to show one possible slice.

3. What other types of shapes can be drawn as a slice in a pyramid?

I can draw quadrilateral and pentagonal slices in a pyramid. I cannot draw a slice with more than five sides because there are only five faces on a pyramid.

EUREKA
MATH

1. Draw a slice into the right rectangular prism at an angle in the form of the provided shape, and draw each slice as a 2D shape.

| **Slice made in the prism** | **Slice as a 2D shape** |

a. A triangle

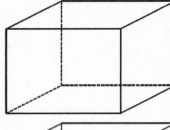

b. A quadrilateral

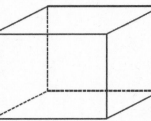

c. A pentagon

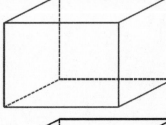

d. A hexagon

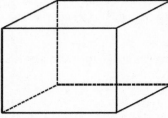

2. Draw slices at an angle in the form of each given shape into each right rectangular pyramid, and draw each slice as a 2D shape.

	Slice made in the pyramid	**Slice as a 2D shape**

a. A triangle

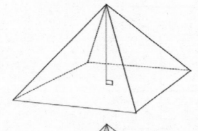

b. A quadrilateral

c. A pentagon

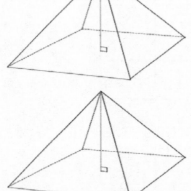

3. Why is it not possible to draw a slice in the shape of a hexagon for a right rectangular pyramid?

4. If the slicing plane meets every face of a right rectangular prism, then the slice is a hexagonal region. What can you say about opposite sides of the hexagon?

5. Draw a right rectangular prism so that rectangles $ABCD$ and $A'B'C'D'$ are base faces. The line segments AA', BB', CC', and DD' are edges of the lateral faces.

a. A slicing plane meets the prism so that vertices A, B, C, and D lie on one side of the plane, and vertices A', B', C', and D' lie on the other side. Based on the slice's position, what other information can be concluded about the slice?

b. A slicing plane meets the prism so that vertices A, B, C, and B' are on one side of the plane, and vertices A', D', C', and D are on the other side. What other information can be concluded about the slice based on its position?

EUREKA
MATH

Example 1

If slices parallel to the tabletop (with height a whole number of units from the tabletop) were taken of this figure, then what would each slice look like?

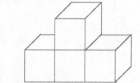

Example 2

If slices parallel to the tabletop were taken of this figure, then what would each slice look like?

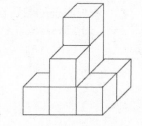

Exercise 1

Based on the level slices you determined in Example 2, how many unit cubes are in the figure?

Exercise 2

a. If slices parallel to the tabletop were taken of this figure, then what would each slice look like?

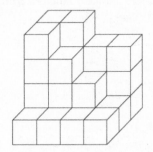

EUREKA
MATH®

b. Given the level slices in the figure, how many unit cubes are in the figure?

Given the level slices in the figure, how many unit cubes are in the figure?

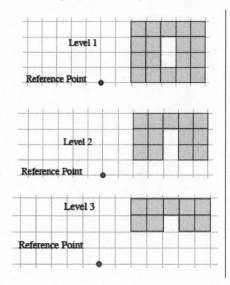

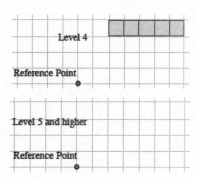

Exercise 3

Sketch your own three-dimensional figure made from cubes and the slices of your figure. Explain how the slices relate to the figure.

Name _____   Date _____

1. The following three-dimensional figure is built on a tabletop. If slices parallel to the tabletop are taken of this figure, then what would each slice look like?

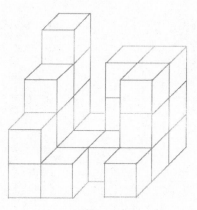

2. Given the level slices in the figure, how many cubes are in the figure?

In the given three-dimensional figures, unit cubes are stacked exactly on top of each other on a tabletop. Each block is either visible or below a visible block.

1.

 a. The following three-dimensional figure is built on a tabletop. If slices parallel to the tabletop are taken of this figure, then what would each slice look like?

Level 1

Reference Point

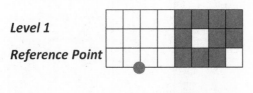

Level 2

Reference Point

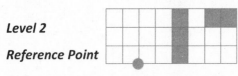

Level 3

Reference Point

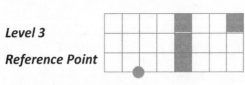

Each reference point shows the cubes that exist in each layer of the figure.

Level 4 and higher

Reference Point

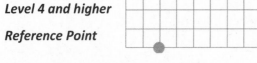

 b. Given the level slices in the figure, how many cubes are in the figure?

   *Level 1: There are **10** cubes between Level 0 and Level 1.*

   *Level 2: There are **5** cubes between Level 1 and Level 2.*

   *Level 3: There are **4** cubes between Level 2 and Level 3.*

   *The total number of cubes in the solid is **19**.*

I know the number of unit cubes can be determined by counting the shaded squares in Levels 1 to 3.

2.

a. The following three-dimensional figure is built on a tabletop. If slices parallel to the tabletop are taken of this figure, then what would each slice look like?

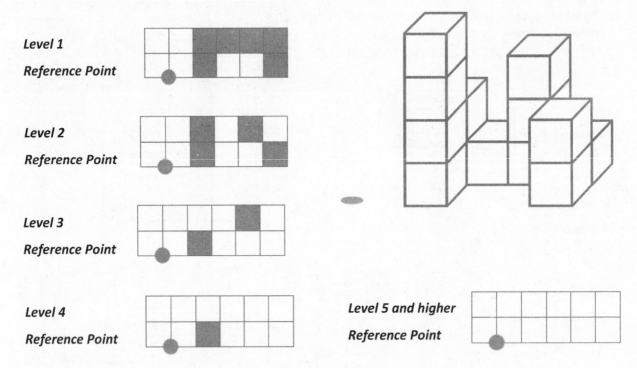

Level 1

Reference Point

Level 2

Reference Point

Level 3

Reference Point

Level 4

Reference Point

Level 5 and higher

Reference Point

b. Given the level slices in the figure, how many cubes are in the figure?

Level 1: There are 6 cubes between Level 0 and Level 1.

Level 2: There are 4 cubes between Level 1 and Level 2.

Level 3: There are 2 cubes between Level 2 and Level 3.

Level 4: There is 1 cube between Level 3 and Level 4.

This time I have four levels of cubes to add together.

The total number of cubes in the solid is 13.

3. When drawing different reference points, why do we not include Level 0?

Level 0 and Level 1 represent the same cubes: Level 0 represents the bottom of these cubes, and Level 1 represents the top of the same cubes. If we showed both Level 0 and Level 1, we would count the same cubes twice.

When I create the reference point drawings, I am showing the top of each cube.

Lesson 19: Understanding Three-Dimensional Figures

EUREKA MATH®

In the given three-dimensional figures, unit cubes are stacked exactly on top of each other on a tabletop. Each block is either visible or below a visible block.

1.

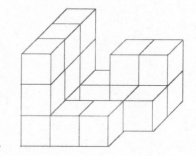

    a.   The following three-dimensional figure is built on a tabletop. If slices parallel to the tabletop are taken of this figure, then what would each slice look like?

    b.   Given the level slices in the figure, how many cubes are in the figure?

2.

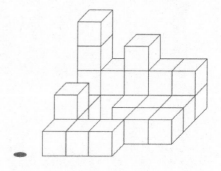

    a.   The following three-dimensional figure is built on a tabletop. If slices parallel to the tabletop are taken of this figure, then what would each slice look like?

    b.   Given the level slices in the figure, how many cubes are in the figure?

3.

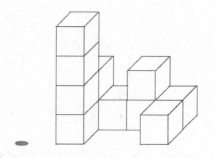

    a.   The following three-dimensional figure is built on a tabletop. If slices parallel to the tabletop are taken of this figure, then what would each slice look like?

    b.   Given the level slices in the figure, how many cubes are in the figure?

4.   John says that we should be including the Level 0 slice when mapping slices. Naya disagrees, saying it is correct to start counting cubes from the Level 1 slice. Who is right?

5.   Draw a three-dimensional figure made from cubes so that each successive layer farther away from the tabletop has one less cube than the layer below it. Use a minimum of three layers. Then draw the slices, and explain the connection between the two.

Opening Exercise

Find the area of each shape based on the provided measurements. Explain how you found each area.

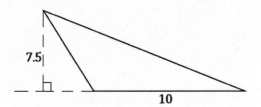

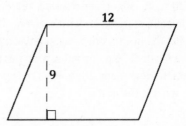

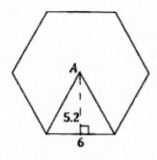

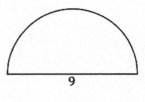

Example 1

A landscape company wants to plant lawn seed. A 20 lb. bag of lawn seed will cover up to 420 sq. ft. of grass and costs $49.98 plus the 8% sales tax. A scale drawing of a rectangular yard is given. The length of the longest side is 100 ft. The house, driveway, sidewalk, garden areas, and utility pad are shaded. The unshaded area has been prepared for planting grass. How many 20 lb. bags of lawn seed should be ordered, and what is the cost?

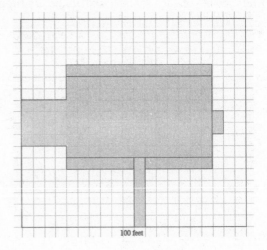

100 feet

Exercise 1

A landscape contractor looks at a scale drawing of a yard and estimates that the area of the home and garage is the same as the area of a rectangle that is 100 ft. × 35 ft. The contractor comes up with 5,500 ft². How close is this estimate?

EUREKA
MATH

Example 2

Ten dartboard targets are being painted as shown in the following figure. The radius of the smallest circle is 3 in., and each successive larger circle is 3 in. more in radius than the circle before it. A can of red paint and a can of white paint is purchased to paint the target. Each 8 oz. can of paint covers 16 ft^2. Is there enough paint of each color to create all ten targets?

Lesson Summary

- One strategy to use when solving area problems with real-world context is to decompose drawings into familiar polygons and circular regions while identifying all relevant measurements.

- Since the area problems involve real-world context, it is important to pay attention to the units needed in each response.

Name _____   Date _____

A homeowner called in a painter to paint the bedroom walls and ceiling. The bedroom is 18 ft. long, 12 ft. wide, and 8 ft. high. The room has two doors each 3 ft. by 7 ft. and three windows each 3 ft. by 5 ft. The doors and windows do not have to be painted. A gallon of paint can cover 300 ft^2. A hired painter claims he will need 4 gallons. Show that the estimate is too high.

1. A farmer has four pieces of unfenced land as shown below in the scale drawing where the dimensions of one side are given. The farmer trades all of the land and $5,000 for 3 acres of similar land that is fenced. If one acre is equal to 43,560 ft^2, how much per square foot for the extra land did the farmer pay rounded to the nearest cent?

$A_1 = \frac{1}{2}(5 \text{ units} \times 4 \text{ units})$

$A_1 = 10 \text{ square units}$

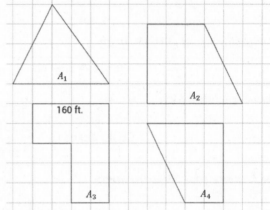

$A_2 = (3 \text{ units} \times 4 \text{ units}) + \frac{1}{2}(2 \text{ units} \times 4 \text{ units})$

$A_2 = 12 \text{ square units} + 4 \text{ square units}$

$A_2 = 16 \text{ square units}$

$A_3 = (4 \text{ units} \times 5 \text{ units}) - (2 \text{ units} \times 3 \text{ units})$

$A_3 = 20 \text{ square units} - 6 \text{ square units}$

$A_3 = 14 \text{ square units}$

> I use composition and decomposition to calculate the areas of the different plots of land.

$A_4 = \frac{1}{2}(2 \text{ units} \times 4 \text{ units}) + (4 \text{ units} \times 2 \text{ units})$

$A_4 = 4 \text{ square units} + 8 \text{ square units}$

$A_4 = 12 \text{ square units}$

The sum of the farmer's four pieces of land:

$A = 10 \text{ square units} + 16 \text{ square units} + 14 \text{ square units} + 12 \text{ square units}$
$A = 52 \text{ square units}$

The sum of the farmer's four pieces of land in square feet:

4 units are 160 feet in length so 1 unit is 40 feet in length. Since each unit is square, each unit is 40 feet in length by 40 feet in width, or 1, 600 ft^2.

The total area of the farmer's pieces of land is 83,200 ft^2 because $52 \times 1,600 = 83,200$.

The sum of the farmer's four pieces of land in acres:

$$83,200 \div 43,560 \approx 1.91$$

> I can use the square feet to determine the number of acres the farmer owns. I use this information to determine the number of extra acres he purchases.

The farmer's four pieces of land total about 1.91 acres.

Extra land purchased with $5,000:

$$3 \text{ acres} - 1.91 \text{ acres} = 1.09 \text{ acres}$$

Extra land in square feet:

$$(1.09 \text{ acres})\left(\frac{43,560 \text{ ft}^2}{1 \text{ acre}} \right) = 47,480.4 \text{ ft}^2$$

> I need to determine the amount of extra square feet before I can determine the cost per square foot.

Price per square foot for extra land:

$$\frac{\$5,000}{47,480.4 \text{ ft}^2} \approx \frac{\$0.11}{\text{ft}^2}$$

The farmer paid about $0.11 per square foot for the extra land.

EUREKA
MATH

2. A stop sign is an octagon with eight equal sides and eight equal angles. The dimensions of the octagon are given below. One side of the octagon is to be painted red. If Derek has enough paint to cover 150 ft², can he paint 50 stop signs? Explain your answer.

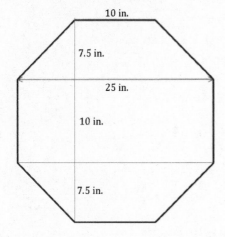

10 in.

7.5 in.

25 in.

10 in.

7.5 in.

I decompose the octagon into three quadrilaterals I am more familiar with.

Area of top trapezoid:

$$A = \frac{1}{2}(10\,\text{in.} + 25\,\text{in.})(7.5\,\text{in.})$$
$$A = 131.25\,\text{in}^2$$

Area of middle rectangle:

$$A = 25\,\text{in.} \times 10\,\text{in.}$$
$$A = 250\,\text{in}^2$$

Area of bottom trapezoid:

$$A = \frac{1}{2}(10\,\text{in.} + 25\,\text{in.})(7.5\,\text{in.})$$
$$A = 131.25\,\text{in}^2$$

Total area of a stop sign in square inches:

$$A = 131.25\,\text{in}^2 + 250\,\text{in}^2 + 131.25\,\text{in}^2$$
$$A = 512.5\,\text{in}^2$$

I need to convert square inches to square feet to determine if Derek has enough paint for 50 stop signs.

Total area of a stop sign in square feet:

$$512.5\,\text{in}^2\left(\frac{1\,\text{ft}^2}{144\,\text{in}^2}\right) \approx 3.56\,\text{ft}^2$$

Total area of 50 stop signs:

$$(3.56\,\text{ft}^2)(50) = 178\,\text{ft}^2$$

Derek does not have enough paint for 50 stop signs because the total area of the stop signs is approximately 178 ft², which is more than 150 ft².

3. A custom home builder is building a new kitchen. The diagram below is of a new kitchen countertop. Approximately how many square feet of counter space is there?

I decompose the counter top into three sections.

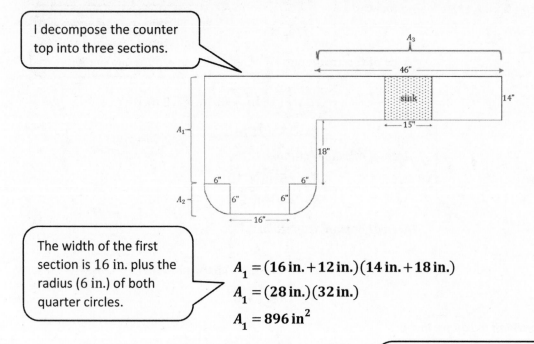

The width of the first section is 16 in. plus the radius (6 in.) of both quarter circles.

$$A_1 = (16\,\text{in.} + 12\,\text{in.})(14\,\text{in.} + 18\,\text{in.})$$
$$A_1 = (28\,\text{in.})(32\,\text{in.})$$
$$A_1 = 896\,\text{in}^2$$

$$A_2 = (16\,\text{in.} \times 6\,\text{in.}) + \frac{1}{4}\pi(6\,\text{in.})^2 + \frac{1}{4}\pi(6\,\text{in.})^2$$
$$A_2 \approx 96\,\text{in}^2 + 28.26\,\text{in}^2 + 28.26\,\text{in}^2$$
$$A_2 \approx 152.52\,\text{in}^2$$

The second section has a rectangle and two quarter circles. I remember the area formula for a quarter circle is $A = \frac{1}{4}\pi r^2$ and I can use 3.14 as an approximate value for π.

I subtract the area of the sink from the area of the third section because the countertop does not cover the sink.

$$A_3 = (46\,\text{in.} \times 14\,\text{in.}) - (15\,\text{in.} \times 14\,\text{in.})$$
$$A_3 = 644\,\text{in}^2 - 210\,\text{in}^2$$
$$A_3 = 434\,\text{in}^2$$

Total area of counter space in square inches:

$$A \approx 896\,\text{in}^2 + 152.52\,\text{in}^2 + 434\,\text{in}^2$$
$$A \approx 1,482.52\,\text{in}^2$$

Total area of counter space in square feet:

$$(1,482.52\,\text{in}^2)\left(\frac{1\,\text{ft}^2}{144\,\text{in}^2}\right) \approx 10.3\,\text{ft}^2$$

To answer the question, I need to convert in^2 to ft^2. I know 12 in. are in 1 ft., which means $144\,\text{in}^2$ are in $1\,\text{ft}^2$.

There is approximately $10.3\,\text{ft}^2$ of counter space.

EUREKA
MATH

1. A farmer has four pieces of unfenced land as shown to the right in the scale drawing where the dimensions of one side are given. The farmer trades all of the land and $10,000 for 8 acres of similar land that is fenced. If one acre is equal to 43,560 ft², how much per square foot for the extra land did the farmer pay rounded to the nearest cent?

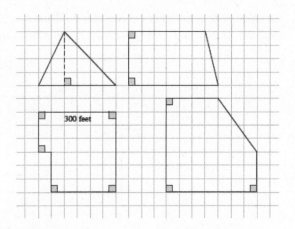

2. An ordinance was passed that required farmers to put a fence around their property. The least expensive fences cost $10 for each foot. Did the farmer save money by moving the farm?

3. A stop sign is an octagon (i.e., a polygon with eight sides) with eight equal sides and eight equal angles. The dimensions of the octagon are given. One side of the stop sign is to be painted red. If Timmy has enough paint to cover 500 ft², can he paint 100 stop signs? Explain your answer.

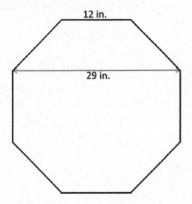

4. The Smith family is renovating a few aspects of their home. The following diagram is of a new kitchen countertop. Approximately how many square feet of counter space is there?

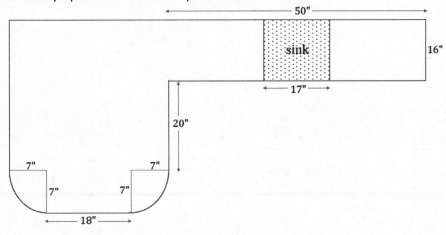

5. In addition to the kitchen renovation, the Smiths are laying down new carpet. Everything but closets, bathrooms, and the kitchen will have new carpet. How much carpeting must be purchased for the home?

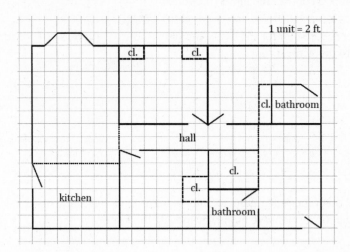

1 unit = 2 ft

6. Jamie wants to wrap a rectangular sheet of paper completely around cans that are $8\frac{1}{2}$ in. high and 4 in. in diameter. She can buy a roll of paper that is $8\frac{1}{2}$ in. wide and 60 ft. long. How many cans will this much paper wrap?

EUREKA
MATH

Opening Exercise

Patty is interested in expanding her backyard garden. Currently, the garden plot has a length of 4 ft. and a width of 3 ft.

    a.   What is the current area of the garden?

Patty plans on extending the length of the plot by 3 ft. and the width by 2 ft.

    b.   What will the new dimensions of the garden be? What will the new area of the garden be?

    c.   Draw a diagram that shows the change in dimension and area of Patty's garden as she expands it. The diagram should show the original garden as well as the expanded garden.

d. Based on your diagram, can the area of the garden be found in a way other than by multiplying the length by the width?

e. Based on your diagram, how would the area of the original garden change if only the length increased by 3 ft.? By how much would the area increase?

f. How would the area of the original garden change if only the width increased by 2 ft.? By how much would the area increase?

g. Complete the following table with the numeric expression, area, and increase in area for each change in the dimensions of the garden.

Dimensions of the Garden	Numeric Expression for the Area of the Garden	Area of the Garden	Increase in Area of the Garden
The original garden with length of 4 ft. and width of 3 ft.			
The original garden with length extended by 3 ft. and width extended by 2 ft.			
The original garden with only the length extended by 3 ft.			
The original garden with only the width extended by 2 ft.			

h. Will the increase in both the length and width by 3 ft. and 2 ft., respectively, mean that the original area will increase strictly by the areas found in parts (e) and (f)? If the area is increasing by more than the areas found in parts (e) and (f), explain what accounts for the additional increase.

Lesson 21:   Mathematical Area Problems

EUREKA
MATH®

Example 1

Examine the change in dimension and area of the following square as it increases by 2 units from a side length of 4 units to a new side length of 6 units. Observe the way the area is calculated for the new square. The lengths are given in units, and the areas of the rectangles and squares are given in units squared.

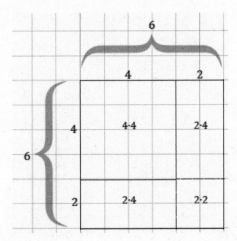

a. Based on the example above, draw a diagram for a square with a side length of 3 units that is increasing by 2 units. Show the area calculation for the larger square in the same way as in the example.

b. Draw a diagram for a square with a side length of 5 units that is increased by 3 units. Show the area calculation for the larger square in the same way as in the example.

c. Generalize the pattern for the area calculation of a square that has an increase in dimension. Let the length of the original square be a units and the increase in length be b units. Use the diagram below to guide your work.

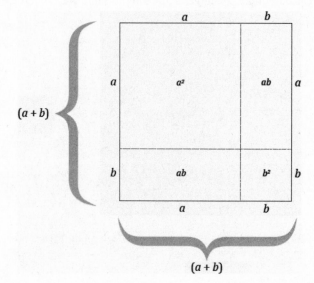

Lesson 21:   Mathematical Area Problems

EUREKA MATH

Example 2

Bobby draws a square that is 10 units by 10 units. He increases the length by x units and the width by 2 units.

a. Draw a diagram that models this scenario.

b. Assume the area of the large rectangle is 156 units^2. Find the value of x.

Example 3

The dimensions of a square with a side length of x units are increased. In this figure, the indicated lengths are given in units, and the indicated areas are given in units^2.

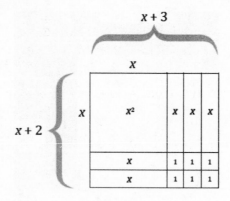

a. What are the dimensions of the large rectangle in the figure?

b. Use the expressions in your response from part (a) to write an equation for the area of the large rectangle, where A represents area.

c. Use the areas of the sections within the diagram to express the area of the large rectangle.

d. What can be concluded from parts (b) and (c)?

e. Explain how the expressions $(x + 2)(x + 3)$ and $x^2 + 3x + 2x + 6$ differ within the context of the area of the figure.

EUREKA
MATH®

Lesson Summary

- The properties of area are limited to positive numbers for lengths and areas.
- The properties of area do support why the properties of operations are true.

Name _____   Date _____

1. Create an area model to represent this product: $(x + 4)(x + 2)$.

2. Write two different expressions that represent the area.

3. Explain how each expression represents different information about the situation.

4. Show that the two expressions are equal using the distributive property.

1. In class, we generalized that $(a + b)^2 = a^2 + 2ab + b^2$. Use these results to evaluate the following expressions by writing $51 = 50 + 1$ and so on.

   a. Evaluate 51^2.

   > I can decompose 51 to the expression $50 + 1$. Therefore, 50 represents a in the general equation, and 1 represents b in the general equation.

   $$51^2 = (50 + 1)^2$$
   $$= 50^2 + 2(50 \cdot 1) + 1^2$$
   $$= 2{,}500 + 100 + 1$$
   $$= 2{,}601$$

   b. Evaluate 201^2.

   $$201^2 = (200 + 1)^2$$
   $$= 200^2 + 2(200 \cdot 1) + 1^2$$
   $$= 40{,}000 + 400 + 1$$
   $$= 40{,}401$$

   > I refer back to the general equation and substitute 200 for a and 1 for b. After substitution, I follow order of operations to simplify the expression.

   c. We can also generalize $(a - b) = a^2 - 2ab + b^2$. Use these results to evaluate the following expression by writing $99 = 100 - 1$, etc.

   Evaluate 99^2.

   $$99^2 = (100 - 1)^2$$
   $$= 100^2 - 2(100 \cdot 1) + 1^2$$
   $$= 10{,}000 - 200 + 1$$
   $$= 9{,}801$$

   > I recognize that I can decompose 99 into a subtraction expression. Therefore, I need to use the general rule for $(a - b)^2$ instead of $(a + b)^2$.

2. Use your knowledge that $a^2 - b^2 = (a - b)(a - b)$ to explain why:

   a. $40^2 - 10^2 = (30)(50)$.

   $$40^2 - 10^2 = (40 - 10)(40 + 10) = (30)(50)$$

   > I can use the general equation to prove that the given number sentence is true. The first term, 40, represents a, and the second term, 10, represents b.

   b. $87^2 - 45^2 = (42)(132)$.

   $$87^2 - 45^2 = (87 - 45)(87 + 45) = (42)(132)$$

3. Create a model for the product. Use the area model to write an equivalent expression that represents the area.

$(x + 2)(x + 3)$

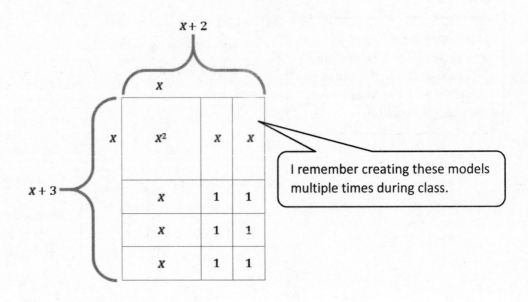

$x + 2$

I remember creating these models multiple times during class.

$x + 3$

$x^2 + 2x + 3x + 6$

4. Use the distributive property to multiply the following expressions.

a. $(3 + 9)(4 + 8)$

$$(3 + 9)(4 + 8) = (3 + 9) \cdot 4 + (3 + 9) \cdot 8$$
$$= 3(4) + 9(4) + 3(8) + 9(8)$$
$$= 12 + 36 + 24 + 72$$
$$= 144$$

I multiply each value in the first set of parentheses by each of the values in the second set of parentheses and then simplify.

b. $(h - 5)(h + 5)$

$$(h - 5)(h + 5) = (h - 5) \cdot h + (h - 5) \cdot 5$$
$$= h(h) - 5(h) + h(5) - 5(5)$$
$$= h^2 - 5h + 5h - 25$$
$$= h^2 - 25$$

I know $-5h$ and $5h$ are opposites, so their sum is 0. Therefore, there are only two terms in the product.

There is a variable present in this expression, which means that I collect like terms.

EUREKA MATH

1. A square with a side length of a units is decreased by b units in both length and width.

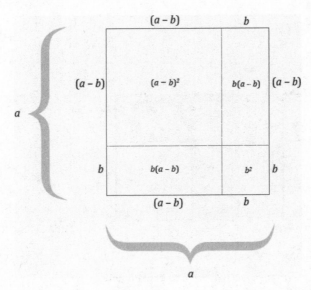

   Use the diagram to express $(a-b)^2$ in terms of the other a^2, ab, and b^2 by filling in the blanks below:

   $(a-b)^2 = a^2 - b(a-b) - b(a-b) - b^2$

   $\quad = a^2 - \underline{\quad} + \underline{\quad} - \underline{\quad} + \underline{\quad} - b^2$

   $\quad = a^2 - 2ab + \underline{\quad} - b^2$

   $\quad = \underline{\hspace{4cm}}$

2. In Example 3, part (c), we generalized that $(a+b)^2 = a^2 + 2ab + b^2$. Use these results to evaluate the following expressions by writing $1{,}001 = 1{,}000 + 1$.

   a. Evaluate 101^2.

   b. Evaluate $1{,}001^2$.

   c. Evaluate 21^2.

3. Use the results of Problem 1 to evaluate 999^2 by writing $999 = 1{,}000 - 1$.

4. The figures below show that $8^2 - 5^2$ is equal to $(8 - 5)(8 + 5)$.

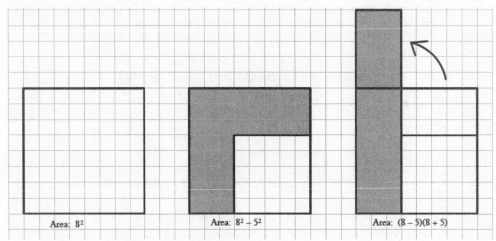

Area: 8^2 Area: $8^2 - 5^2$ Area: $(8 - 5)(8 + 5)$

a. Create a drawing to show that $a^2 - b^2 = (a - b)(a + b)$.

b. Use the result in part (a), $a^2 - b^2 = (a - b)(a + b)$, to explain why:

   i. $35^2 - 5^2 = (30)(40)$.

   ii. $21^2 - 18^2 = (3)(39)$.

   iii. $104^2 - 63^2 = (41)(167)$.

c. Use the fact that $35^2 = (30)(40) + 5^2 = 1,225$ to create a way to mentally square any two-digit number ending in 5.

5. Create an area model for each product. Use the area model to write an equivalent expression that represents the area.

a. $(x + 1)(x + 4) =$

b. $(x + 5)(x + 2) =$

c. Based on the context of the area model, how do the expressions provided in parts (a) and (b) differ from the equivalent expression answers you found for each?

6. Use the distributive property to multiply the following expressions.

a. $(2 + 6)(2 + 4)$

b. $(x + 6)(x + 4)$; draw a figure that models this multiplication problem.

c. $(10 + 7)(10 + 7)$

d. $(a + 7)(a + 7)$

e. $(5 - 3)(5 + 3)$

f. $(x - 3)(x + 3)$

EUREKA MATH

Example 1

a. The circle to the right has a diameter of 12 cm. Calculate the area of the shaded region.

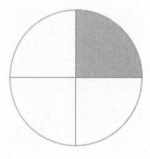

b. Sasha, Barry, and Kyra wrote three different expressions for the area of the shaded region. Describe what each student was thinking about the problem based on his or her expression.

Sasha's expression: $\frac{1}{4}\pi(6^2)$

Barry's expression: $\pi(6^2) - \frac{3}{4}\pi(6^2)$

Kyra's expression: $\frac{1}{2}\left(\frac{1}{2}\pi(6^2)\right)$

Exercise 1

a. Find the area of the shaded region of the circle to the right.

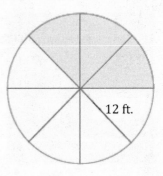

12 ft.

b. Explain how the expression you used represents the area of the shaded region.

Exercise 2

Calculate the area of the figure below that consists of a rectangle and two quarter circles, each with the same radius. Leave your answer in terms of pi.

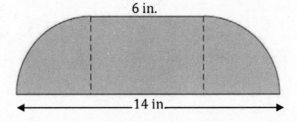

6 in.

14 in.

EUREKA
MATH

Example 2

The square in this figure has a side length of 14 inches. The radius of the quarter circle is 7 inches.

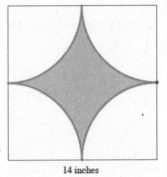

14 inches

a.   Estimate the shaded area.

b.   What is the exact area of the shaded region?

c.   What is the approximate area using $\pi \approx \dfrac{22}{7}$?

Exercise 3

The vertices A and B of rectangle $ABCD$ are centers of circles each with a radius of 5 inches.

a. Find the exact area of the shaded region.

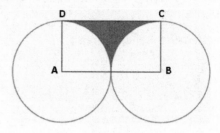

b. Find the approximate area using $\pi \approx \dfrac{22}{7}$.

c. Find the area to the nearest hundredth using the π key on your calculator.

EUREKA
MATH

Exercise 4

The diameter of the circle is 12 in. Write and explain a numerical expression that represents the area of the shaded region.

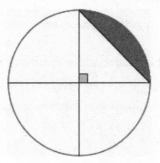

Lesson Summary

To calculate composite figures with circular regions:

- Identify relevant geometric areas (such as rectangles or squares) that are part of a figure with a circular region.

- Determine which areas should be subtracted or added based on their positions in the diagram.

- Answer the question, noting if the exact or approximate area is to be found.

Name _____ Date _____

A circle with a 10 cm radius is cut into a half circle and two quarter circles. The three circular arcs bound the region below.

  a. Write and explain a numerical expression that represents the area.

  b. Then, find the area of the figure.

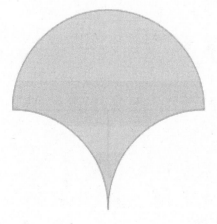

1. A circle with center O has an area of 117 in². Find the area of the unshaded region.

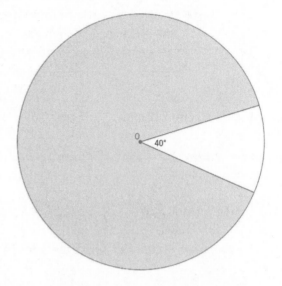

> I know the entire circle represents $360°$, which means $40°$ represents $\frac{40}{360}$, or $\frac{1}{9}$, of the circle. Therefore, the area of the unshaded region is $\frac{1}{9}$ of the area of the entire circle.

$$A = \frac{1}{9}(117 \text{ in}^2) = 13 \text{ in}^2$$

The area of the unshaded region is 13 in².

2. The following region is bounded by the arcs of two quarter circles, each with a radius of 7 cm, and by line segments 10 cm in length. The region on the right shows a rectangle with dimensions 7 cm by 10 cm. Show that both shaded regions have equal areas.

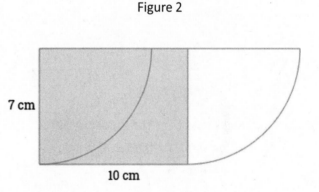

Figure 1   Figure 2

7 cm   10 cm

7 cm

10 cm

7 cm

10 cm

> For Figure 1, I subtract the area of the quarter circle from the area of the rectangle and then add the area of the extra quarter circle.

> The shaded region in Figure 2 is a rectangle, which means I find the product of the length and width to calculate the area.

$$A = \left(10\,\text{cm} \times 7\,\text{cm} - \frac{1}{4}\pi(7\,\text{cm})^2 \right) + \frac{1}{4}\pi(7\,\text{cm})^2$$

$$A = 70\,\text{cm}^2 - \frac{49\pi}{4}\,\text{cm}^2 + \frac{49\pi}{4}\,\text{cm}^2$$

$$A = 70\,\text{cm}^2$$

$$A = 10\,\text{cm} \times 7\,\text{cm}$$

$$A = 70\,\text{cm}^2$$

3. The diameters of four half circles are sides of a square with a side length of 6 cm.

6 cm

Figure 1

> Figure 2 isolates one quarter of Figure 1.

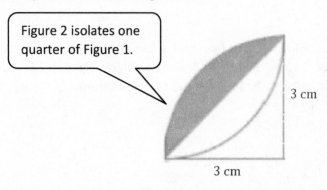

3 cm

3 cm

Figure 2
(Not drawn to scale)

a. Find the exact area of the shaded region.

   Area of the shaded region in Figure 2:

$$\frac{1}{4}\pi(3\text{ cm})^2 - \frac{1}{2}(3\text{ cm} \times 3\text{ cm})$$

$$\frac{9}{4}\pi\text{ cm}^2 - 4.5\text{ cm}^2$$

> To calculate the shaded region in Figure 2, I calculate the area of a quarter circle and subtract the area of a right triangle.

   Total shaded area:

$$8\left(\frac{9}{4}\pi\text{ cm}^2 - 4.5\text{ cm}^2\right)$$

$$18\pi\text{ cm}^2 - 36\text{ cm}^2$$

> There are 8 identical shaded regions, so I multiply the area by 8. To calculate the exact area, I leave π in the answer.

The exact area of the shaded region is 18π cm^2 − 36 cm^2.

EUREKA MATH

b. Find the approximate area using $\pi \approx \frac{22}{7}$.

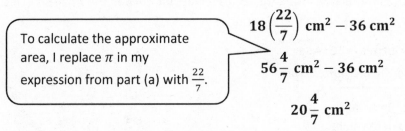

To calculate the approximate area, I replace π in my expression from part (a) with $\frac{22}{7}$.

$$18\left(\frac{22}{7}\right) \text{ cm}^2 - 36 \text{ cm}^2$$

$$56\frac{4}{7} \text{ cm}^2 - 36 \text{ cm}^2$$

$$20\frac{4}{7} \text{ cm}^2$$

The approximate area of the shaded region is $20\frac{4}{7}$ cm^2.

4. A square with a side length 8 inches is shown below, along with a quarter circle (with a side of the square as its radius) and two half circles (with diameters that are sides of the square). Write and explain a numerical expression that represents the exact area of the shaded region in the figure.

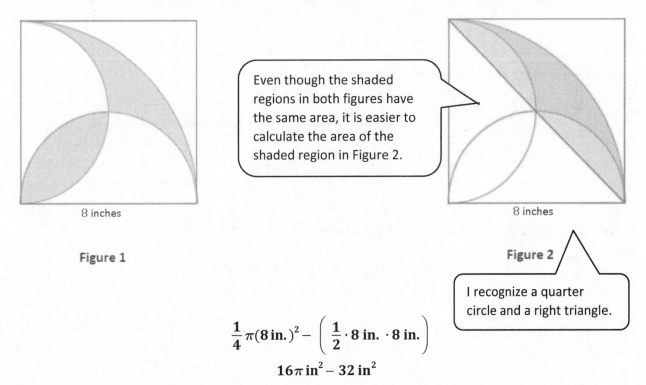

8 inches

Even though the shaded regions in both figures have the same area, it is easier to calculate the area of the shaded region in Figure 2.

8 inches

Figure 1

Figure 2

I recognize a quarter circle and a right triangle.

$$\frac{1}{4}\pi(8 \text{ in.})^2 - \left(\frac{1}{2} \cdot 8 \text{ in.} \cdot 8 \text{ in.} \right)$$

$$16\pi \text{ in}^2 - 32 \text{ in}^2$$

The shaded area in each figure has the same area. This area can be found by subtracting the area of a right triangle with leg lengths of 8 in. from the area of the quarter circle with a radius of 8 in.

5. Four circles have centers on segment YZ. The diameters of the circles are in the ratio $5:2:2:1$. If the area of the largest circle is 100 ft^2, find the area inside the largest circle but outside the smaller circles.

The ratio of the areas of the four circles is $25:4:4:1$.

Let x represent the area of one of the medium circles in ft^2:

> I can use the ratio of the areas and the given area to determine the area of the medium circles.

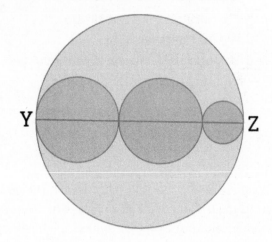

$$\frac{4}{25} = \frac{x}{100}$$

$$(100)\left(\frac{4}{25}\right) = (100)\left(\frac{x}{100}\right)$$

$$16 = x$$

Let y represent the area of the smaller circle in ft^2:

$$\frac{1}{25} = \frac{y}{100}$$

$$(100)\left(\frac{1}{25}\right) = (100)\left(\frac{y}{100}\right)$$

$$4 = y$$

> Once I know the area of the three smaller circles, I can subtract these areas from the given area of the largest circle.

Area inside the largest circle but outside the smaller three circles:

$$A = 100 \text{ ft}^2 - 16 \text{ ft}^2 - 16 \text{ ft}^2 - 4 \text{ ft}^2$$

$$A = 64 \text{ ft}^2$$

The area inside the largest circle but outside the three smaller circles is 64 ft^2.

EUREKA MATH

1. A circle with center O has an area of 96 in^2. Find the area of the shaded region.

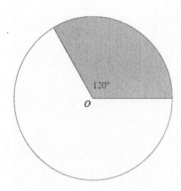

120°

O

Peyton's Solution

$$A = \frac{1}{3}\left(96 \text{ in}^2\right) = 32 \text{ in}^2$$

Monte's Solution

$$A = \frac{96}{120} \text{ in}^2 = 0.8 \text{ in}^2$$

Which person solved the problem correctly? Explain your reasoning.

2. The following region is bounded by the arcs of two quarter circles, each with a radius of 4 cm, and by line segments 6 cm in length. The region on the right shows a rectangle with dimensions 4 cm by 6 cm. Show that both shaded regions have equal areas.

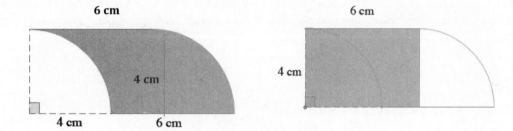

6 cm   6 cm

4 cm   4 cm

4 cm   6 cm

3. A square is inscribed in a paper disc (i.e., a circular piece of paper) with a radius of 8 cm. The paper disc is red on the front and white on the back. Two edges of the circle are folded over. Write and explain a numerical expression that represents the area of the figure. Then, find the area of the figure.

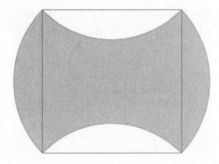

4. The diameters of four half circles are sides of a square with a side length of 7 cm.

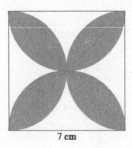

7 cm

a. Find the exact area of the shaded region.

b. Find the approximate area using $\pi \approx \dfrac{22}{7}$.

c. Find the area using the π button on your calculator and rounding to the nearest thousandth.

5. A square with a side length of 14 inches is shown below, along with a quarter circle (with a side of the square as its radius) and two half circles (with diameters that are sides of the square). Write and explain a numerical expression that represents the area of the figure.

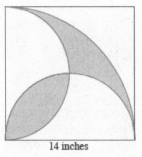

14 inches

6. Three circles have centers on segment AB. The diameters of the circles are in the ratio $3: 2: 1$. If the area of the largest circle is 36 ft^2, find the area inside the largest circle but outside the smaller two circles.

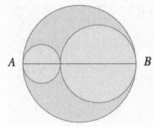

A B

EUREKA
MATH

7.   A square with a side length of 4 ft. is shown, along with a diagonal, a quarter circle (with a side of the square as its radius), and a half circle (with a side of the square as its diameter). Find the exact, combined area of regions I and II.

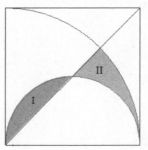

Opening Exercise

Calculate the surface area of the square pyramid.

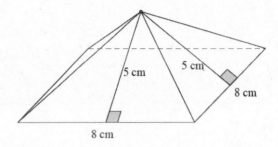

Example 1

a. Calculate the surface area of the rectangular prism.

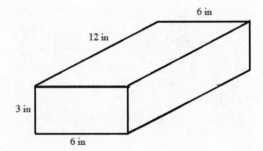

b.   Imagine that a piece of the rectangular prism is removed. Determine the surface area of both pieces.

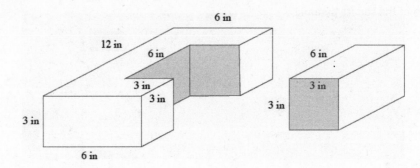

c.   How is the surface area in part (a) related to the surface area in part (b)?

EUREKA
MATH

Exercises

Determine the surface area of the right prisms.

1.

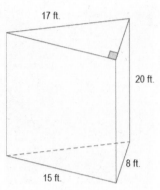

2.

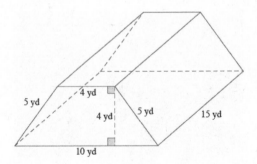

3.

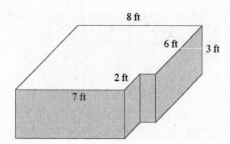

4.

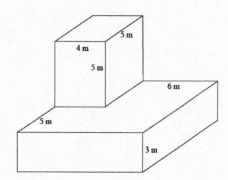

Lesson 23:   Surface Area

EUREKA
MATH®

5.

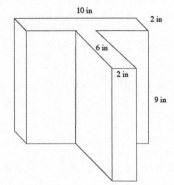

Lesson Summary

To determine the surface area of right prisms that are composite figures or missing sections, determine the area of each lateral face and the two base faces, and then add the areas of all the faces together.

Lesson 23: Surface Area

Name _____   Date _____

Determine and explain how to find the surface area of the following right prisms.

1.

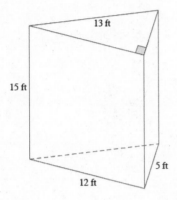

13 ft

15 ft

12 ft

5 ft

2.

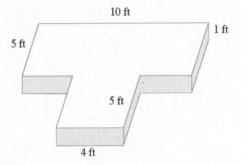

10 ft

1 ft

5 ft

5 ft

4 ft

EUREKA MATH

> To calculate the surface area of a figure, I find the area of each face and then calculate the sum of these areas.

> Even though the top and bottom of the figure look different, they cover the same area.

Determine the surface area of the figures.

1.

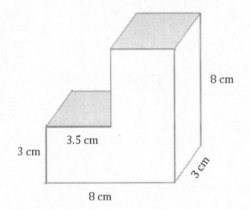

Area of the top and bottom: $2(8 \text{ cm} \times 3 \text{ cm}) = 48 \text{ cm}^2$

Area of left and right sides: $2(3 \text{ cm} \times 8 \text{ cm}) = 48 \text{ cm}^2$

Area of front and back:
$2(8 \text{ cm} \times 3 \text{ cm}) + 2(4.5 \text{ cm} \times 5 \text{ cm}) = 93 \text{ cm}^2$

> I decompose the front and back faces into two rectangles in order to calculate the area.

Total surface area: $48 \text{ cm}^2 + 48 \text{ cm}^2 + 93 \text{ cm}^2 = 189 \text{ cm}^2$

2.

> The bottom of the top prism is not part of the surface area.

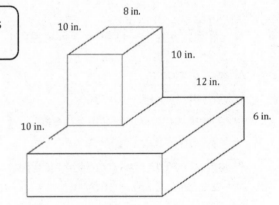

Surface area of top prism:

Area of top: $10 \text{ in.} \times 8 \text{ in.} = 80 \text{ in}^2$

Area of front and back sides: $2(8 \text{ in.} \times 10 \text{ in.}) = 160 \text{ in}^2$

Area of left and right sides: $2(10 \text{ in.} \times 10 \text{ in.}) = 200 \text{ in}^2$

Surface area of bottom prism:

Area of top: $(20 \text{ in.} \times 20 \text{ in.}) - 80 \text{ in}^2 = 320 \text{ in}^2$

Area of bottom: $20 \text{ in.} \times 20 \text{ in.} = 400 \text{ in}^2$

Area of front and back sides: $2(6 \text{ in.} \times 20 \text{ in.}) = 240 \text{ in}^2$

Area of left and right sides: $2(6 \text{ in.} \times 20 \text{ in.}) = 240 \text{ in}^2$

> I subtract 80 in^2 from the area of the top of the bottom prism because this is the area that the top and bottom prisms overlap.

Total surface area: $80 \text{ in}^2 + 160 \text{ in}^2 + 200 \text{ in}^2 + 320 \text{ in}^2 + 400 \text{ in}^2 + 240 \text{ in}^2 + 240 \text{ in}^2 = 1,640 \text{ in}^2$

3.

> I know that the prism has a rectangular base and four triangular faces. I remember that the area formula for a triangle is $A = \frac{1}{2}bh$.

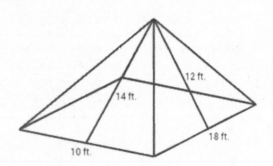

Area of front and back: $2\left(\frac{1}{2}(10\text{ ft.} \times 14\text{ ft.})\right) = 140\text{ ft}^2$

Area of left and right: $2\left(\frac{1}{2}(18\text{ ft.} \times 12\text{ ft.})\right) = 216\text{ ft}^2$

Area of the base: $10\text{ ft.} \times 18\text{ ft.} = 180\text{ ft}^2$

Total surface area: $140\text{ ft}^2 + 216\text{ ft}^2 + 180\text{ ft}^2 = 536\text{ ft}^2$

4.

Area of the front and back:

$$2\left(\frac{1}{2}(7\text{ cm} \times 24\text{ cm})\right) = 168\text{ cm}^2$$

> I know the triangular prism has two triangular bases and three rectangular faces. I need to find the area of each of these five faces in order to calculate the surface area.

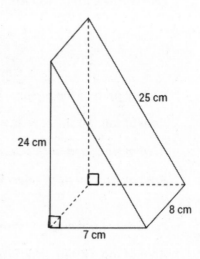

Area of bottom:

$7\text{ cm} \times 8\text{ cm} = 56\text{ cm}^2$

Area that can be seen from left side:

$24\text{ cm} \times 8\text{ cm} = 192\text{ cm}^2$

Area that can be seen from the right side:

$25\text{ cm} \times 8\text{ cm} = 200\text{ cm}^2$

Total surface area: $168\text{ cm}^2 + 56\text{ cm}^2 + 192\text{ cm}^2 + 200\text{ cm}^2 = 616\text{ cm}^2$

EUREKA
MATH

Determine the surface area of the figures.

1.

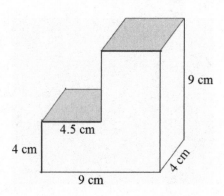

2.

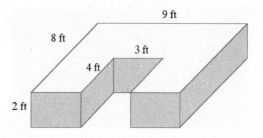

3.

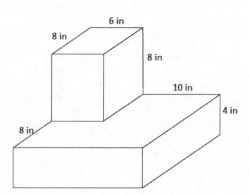

4.

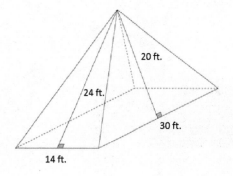

5.

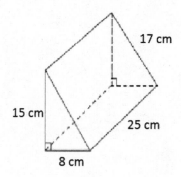

Example 1

Determine the surface area of the image.

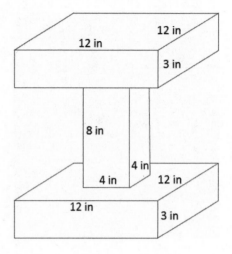

Example 2

a. Determine the surface area of the cube.

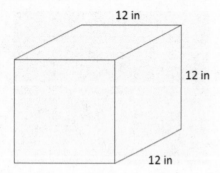

b. A square hole with a side length of 4 inches is cut through the cube. Determine the new surface area.

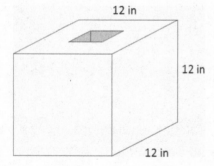

EUREKA
MATH

Example 3

A right rectangular pyramid has a square base with a side length of 10 inches. The surface area of the pyramid is 260 in^2. Find the height of the four lateral triangular faces.

Exercises

Determine the surface area of each figure. Assume all faces are rectangles unless it is indicated otherwise.

1.

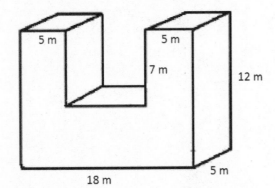

2. In addition to your calculation, explain how the surface area of the following figure was determined.

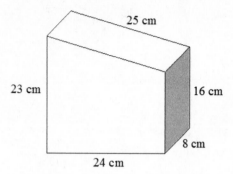

3.

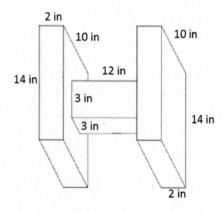

EUREKA
MATH®

4. In addition to your calculation, explain how the surface area was determined.

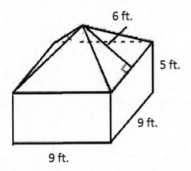

5. A hexagonal prism has the following base and has a height of 8 units. Determine the surface area of the prism.

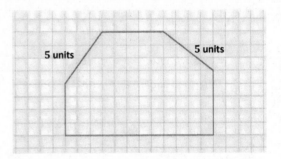

6. Determine the surface area of each figure.

a.

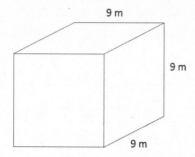

b. A cube with a square hole with 3 m side lengths has been cut through the cube.

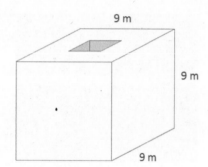

c. A second square hole with 3 m side lengths has been cut through the cube.

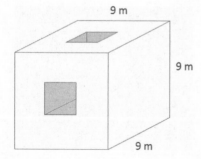

Lesson 24:   Surface Area

EUREKA
MATH®

7. The figure below shows 28 cubes with an edge length of 1 unit. Determine the surface area.

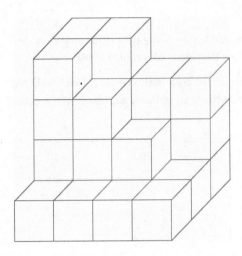

8. The base rectangle of a right rectangular prism is 4 ft. × 6 ft. The surface area is 288 ft^2. Find the height. Let h be the height in feet.

EUREKA MATH®

Name _____   Date _____

Determine the surface area of the right rectangular prism after the two square holes have been cut. Explain how you determined the surface area.

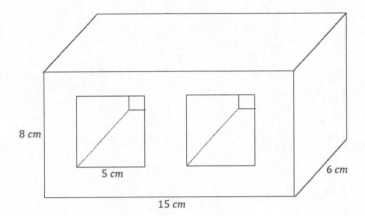

Determine the surface area of each figure.

1.

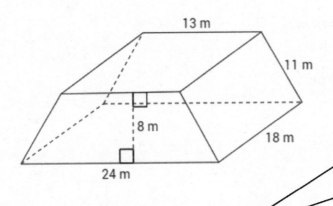

I know that trapezoidal prisms have two bases that are trapezoids. The area formula for a trapezoid is $A = \frac{1}{2}(b_1 + b_2)h$, where b_1 and b_2 represent the lengths of the two bases.

Area of front and back: $2\left(\frac{1}{2}(13\,\text{m} + 24\,\text{m})8\,\text{m}\right) = 296\,\text{m}^2$

Area of top: $13\,\text{m} \times 18\,\text{m} = 234\,\text{m}^2$

Area of left and right sides: $2(11\,\text{m} \times 18\,\text{m}) = 396\,\text{m}^2$

The other four faces are all rectangles.

Area of bottom: $24\,\text{m} \times 18\,\text{m} = 432\,\text{m}^2$

Total surface area: $296\,\text{m}^2 + 234\,\text{m}^2 + 396\,\text{m}^2 + 432\,\text{m}^2 = 1,358\,\text{m}^2$

2. Determine the surface area after two square holes with a side length of 3 m are cut through the solid figure composed of two rectangular prisms.

Surface area of the top prism before the hole is cut:

Area of top: $5 \text{ m} \times 6 \text{ m} = 30 \text{ m}^2$

Area of front and back: $2(5 \text{ m} \times 6 \text{ m}) = 60 \text{ m}^2$

Area of sides: $2(6 \text{ m} \times 6 \text{ m}) = 72 \text{ m}^2$

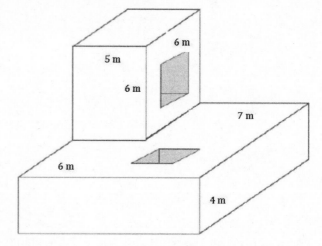

> I first find the surface area of both prisms before the holes are cut.

Surface area of the bottom prism before the hole is cut:

Area of top: $12 \text{ m} \times 12 \text{ m} - 30 \text{ m}^2 = 114 \text{ m}^2$

Area of bottom: $12 \text{ m} \times 12 \text{ m} = 144 \text{ m}^2$

Area of front and back: $2(12 \text{ m} \times 4 \text{ m}) = 96 \text{ m}^2$

Area of sides: $2(12 \text{ m} \times 4 \text{ m}) = 96 \text{ m}^2$

> Each of the four faces inside each hole are rectangles. The width of all eight rectangles is 3 m. The length of the rectangles in the top hole is 5 m. The length of the rectangles in the bottom hole is the same as the height of the bottom prism, or 4 m.

Surface area of holes: $4(3 \text{ m} \times 5 \text{ m}) + 4(3 \text{ m} \times 4 \text{ m}) = 108 \text{ m}^2$

> I add all the areas together, but I also have to subtract the places where holes are cut. Each hole is a 3 m by 3 m square, which means the area cut is 9 m². There are four places this area is cut out of the original prism, which means I subtract an area of 36 m² from the total surface area.

Total surface area:

$30 \text{ m}^2 + 60 \text{ m}^2 + 72 \text{ m}^2 + 114 \text{ m}^2 + 144 \text{ m}^2 + 96 \text{ m}^2 + 96 \text{ m}^2 + 108 \text{ m}^2 - 36 \text{ m}^2 = 684 \text{ m}^2$

EUREKA MATH

3. The base of a right prism is shown below. Determine the surface area if the height of the prism is 6 cm. Explain how you determined the surface area.

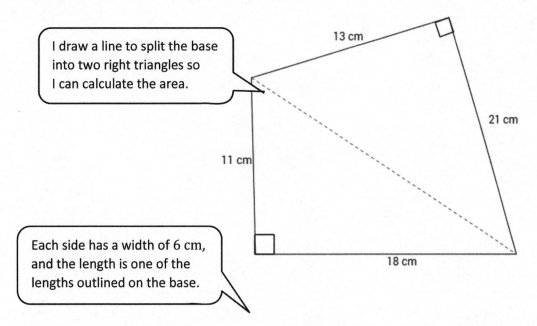

I draw a line to split the base into two right triangles so I can calculate the area.

Each side has a width of 6 cm, and the length is one of the lengths outlined on the base.

Area of sides: $(11 \text{ cm} \times 6 \text{ cm}) + (18 \text{ cm} \times 6 \text{ cm}) + (21 \text{ cm} \times 6 \text{ cm}) + (13 \text{ cm} \times 6 \text{ cm}) = 378 \text{ cm}^2$

Area of bases: $2\left(\frac{1}{2}(11 \text{ cm} \times 18 \text{ cm})\right) + 2\left(\frac{1}{2}(21 \text{ cm} \times 13 \text{ cm})\right) = 471 \text{ cm}^2$

Total surface area: $471 \text{ cm}^2 + 378 \text{ cm}^2 = 849 \text{ cm}^2$

Determine the surface area of each figure.

1. In addition to the calculation of the surface area, describe how you found the surface area.

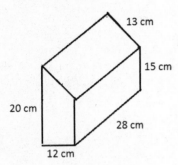

2.

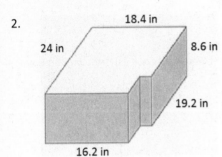

3.

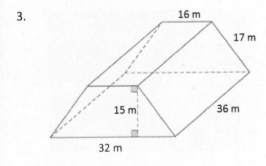

4. Determine the surface area after two square holes with a side length of 2 m are cut through the solid figure composed of two rectangular prisms.

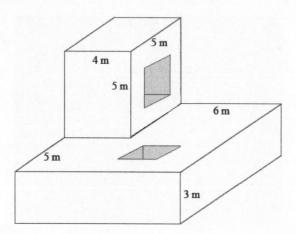

5. The base of a right prism is shown below. Determine the surface area if the height of the prism is 10 cm. Explain how you determined the surface area.

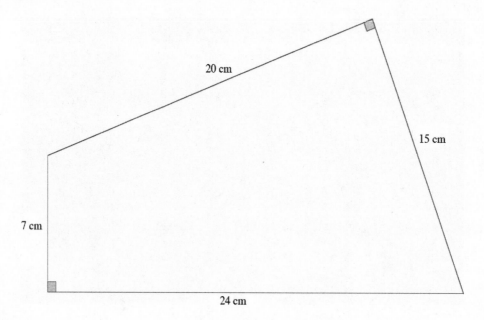

Opening Exercise

Take your copy of the following figure, and cut it into four pieces along the dotted lines. (The vertical line is the altitude, and the horizontal line joins the midpoints of the two sides of the triangle.)

Arrange the four pieces so that they fit together to form a rectangle.

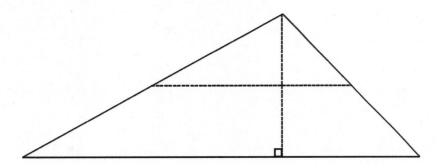

If a prism were formed out of each shape, the original triangle, and your newly rearranged rectangle, and both prisms had the same height, would they have the same volume? Discuss with a partner.

Exercise 1

a. Show that the following figures have equal volumes.

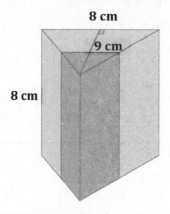

b. How can it be shown that the prisms will have equal volumes without completing the entire calculation?

Calculate the volume of the following prism.

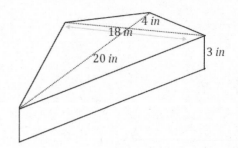

EUREKA
MATH

Example 2

A container is shaped like a right pentagonal prism with an open top. When a cubic foot of water is dumped into the container, the depth of the water is 8 inches. Find the area of the pentagonal base.

Example 3

Two containers are shaped like right triangular prisms, each with the same height. The base area of the larger container is 200% more than the base area of the smaller container. How many times must the smaller container be filled with water and poured into the larger container in order to fill the larger container?

Exercise 2

Two aquariums are shaped like right rectangular prisms. The ratio of the dimensions of the larger aquarium to the dimensions of the smaller aquarium is $3 : 2$.

Addie says the larger aquarium holds 50% more water than the smaller aquarium.

Berry says that the larger aquarium holds 150% more water.

Cathy says that the larger aquarium holds over 200% more water.

Are any of the girls correct? Explain your reasoning.

Lesson 25:   Volume of Right Prisms

Lesson Summary

- The formula for the volume of a prism is $V = Bh$, where B is the area of the base of the prism and h is the height of the prism.

- A base that is neither a rectangle nor a triangle must be decomposed into rectangles and triangles in order to find the area of the base.

Opening Exercise

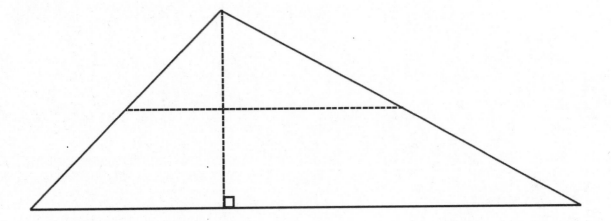

EUREKA
MATH

Name _____   Date _____

Determine the volume of the following prism. Explain how you found the volume.

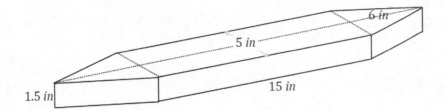

© 2019 Great Minds®. eureka-math.org

1. Two right prism containers each hold 37.5 gallons of water. The height of the first container is 12 inches. The height of the second container is 10 inches. If the area of the base of the first container is 5 ft², find the area of the base of the second container. Explain your reasoning.

Let B represent the area of the base in the second container.

$$12 \times 5 = 10 \times B$$
$$60 = 10B$$
$$\left(\frac{1}{10}\right)(60) = \left(\frac{1}{10}\right)(10B)$$
$$6 = B$$

> I know the two containers have the same volume. Therefore, the product of the area of the base and the height of each container must be equal.

The area of the base of the second container is 6 ft².

> There are three different dimensions in a right rectangular prism: length, width, and height.

2. Two containers are shaped like right rectangular prisms. Each of the larger container's dimensions are 30% more than the smaller container's dimensions. If the smaller container holds 15 gallons when full, how many gallons does the larger container hold? Explain your reasoning.

> Each dimension of the larger container is 1.3 times larger than those of the smaller container because 100% + 30% is 130%.

The volume of the larger container is 1.3³, or 2.197, times larger than the volume of the smaller container.

$$15 \text{ gallons} \times 2.197 = 32.955 \text{ gallons}$$

> The volume of the larger container is 2.197 times bigger than the volume of the smaller container.

The volume of the larger container is 32.955 gallons.

3. An aquarium in the shape of a right rectangular prism has a base area of 40 in^2 and height of 13 in. Currently, the aquarium is only partially filled, and the height of the water is 8 in. A few decorations are added to the bottom of the aquarium, which makes the water rise to the top, completely submerging the decorations but without causing overflow. Find the volume of the decorations.

The height of the water increased **5** *in. because the height increased from* **8** *in.* *to* **13** *in.*

$$40 \text{ in}^2 \times 5 \text{ in.} = 200 \text{ in}^3$$

The volume of the decorations is **200** *in^3.*

> The volume of the decorations will be the same as the change in the volume of the water. I can calculate the change in the volume by multiplying the change in height by the area of the base.

4. A rectangular swimming pool is 20 feet wide and 40 feet long. The pool is 4 feet deep at one end, and 15 feet deep at the other.

    a. Sketch the swimming pool as a right prism.

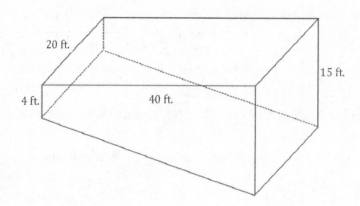

    b. What kind of prism is the swimming pool?

    The swimming pool is a right trapezoidal prism.

> There are two trapezoidal bases in the swimming pool.

EUREKA
MATH®

c. What is the volume of the swimming pool in cubic feet?

> The area of the trapezoidal base is half the sum of the lengths times the height.

$$V = \left(\frac{1}{2}(4\,\text{ft.} + 15\,\text{ft.})(40\,\text{ft.})\right)(20\,\text{ft.})$$

$$V = \left(\frac{1}{2}(19\,\text{ft.})(40\,\text{ft.})\right)(20\,\text{ft.})$$

$$V = \left(380\,\text{ft}^2\right)(19\,\text{ft.})$$

$$V = 7,600\,\text{ft}^3$$

The volume of the swimming pool is $7,600\,\text{ft}^3$**.**

d. How many gallons will the swimming pool hold if each cubic feet of water is about 7.5 gallons?

$$\left(7,600\,\text{ft}^3\right)\left(\frac{7.5\,\text{gallons}}{1\,\text{ft}^3}\right) = 57,000\,\text{gallons}$$

The swimming pool will hold about *57,000 gallons of water.*

1. The pieces in Figure 1 are rearranged and put together to form Figure 2.

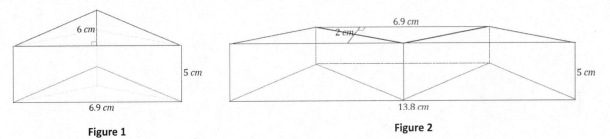

   Figure 1   **Figure 2**

   a. Use the information in Figure 1 to determine the volume of the prism.

   b. Use the information in Figure 2 to determine the volume of the prism.

   c. If we were not told that the pieces of Figure 1 were rearranged to create Figure 2, would it be possible to determine whether the volumes of the prisms were equal without completing the entire calculation for each?

2. Two right prism containers each hold 75 gallons of water. The height of the first container is 20 inches. The of the second container is 30 inches. If the area of the base in the first container is 6 ft², find the area of the base in the second container. Explain your reasoning.

3. Two containers are shaped like right rectangular prisms. Each has the same height, but the base of the larger container is 50% more in each direction. If the smaller container holds 8 gallons when full, how many gallons does the larger container hold? Explain your reasoning.

4. A right prism container with the base area of 4 ft² and height of 5 ft. is filled with water until it is 3 ft. deep. If a solid cube with edge length 1 ft. is dropped to the bottom of the container, how much will the water rise?

5. A right prism container with a base area of 10 ft² and height 9 ft. is filled with water until it is 6 ft. deep. A large boulder is dropped to the bottom of the container, and the water rises to the top, completely submerging the boulder without causing overflow. Find the volume of the boulder.

6. A right prism container with a base area of 8 ft² and height 6 ft. is filled with water until it is 5 ft. deep. A solid cube is dropped to the bottom of the container, and the water rises to the top. Find the length of the cube.

7. A rectangular swimming pool is 30 feet wide and 50 feet long. The pool is 3 feet deep at one end, and 10 feet deep at the other.

   a. Sketch the swimming pool as a right prism.

   b. What kind of right prism is the swimming pool?

   c. What is the volume of the swimming pool in cubic feet?

   d. How many gallons will the swimming pool hold if each cubic feet of water is about 7.5 gallons?

8. A milliliter (mL) has a volume of 1 cm³. A 250 mL measuring cup is filled to 200 mL. A small stone is placed in the measuring cup. The stone is completely submerged, and the water level rises to 250 mL.

   a. What is the volume of the stone in cm³?

   b. Describe a right rectangular prism that has the same volume as the stone.

© 2019 Great Minds®. eureka-math.org

Example 1

Find the volume of the following three-dimensional object composed of two right rectangular prisms.

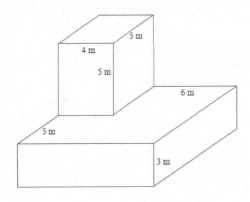

Exercise 1

Find the volume of the following three-dimensional figure composed of two right rectangular prisms.

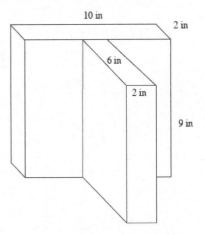

Exercise 2

The right trapezoidal prism is composed of a right rectangular prism joined with a right triangular prism. Find the volume of the right trapezoidal prism shown in the diagram using two different strategies.

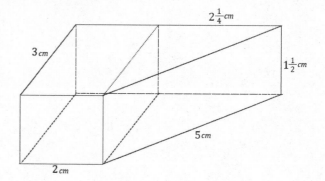

Example 2

Find the volume of the right prism shown in the diagram whose base is the region between two right triangles. Use two different strategies.

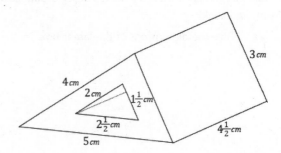

Example 3

A box with a length of 2 ft., a width of 1.5 ft., and a height of 1.25 ft. contains fragile electronic equipment that is packed inside a larger box with three inches of styrofoam cushioning material on each side (above, below, left side, right side, front, and back).

a. Give the dimensions of the larger box.

b. Design styrofoam right rectangular prisms that could be placed around the box to provide the cushioning (i.e., give the dimensions and how many of each size are needed).

c. Find the volume of the styrofoam cushioning material by adding the volumes of the right rectangular prisms in the previous question.

d. Find the volume of the styrofoam cushioning material by computing the difference between the volume of the larger box and the volume of the smaller box.

EUREKA
MATH

> **Lesson Summary**
>
> To find the volume of a three-dimensional composite object, two or more distinct volumes must be added together (if they are joined together) or subtracted from each other (if one is a missing section of the other). There are two strategies to find the volume of a prism:
>
> - Find the area of the base and then multiply times the prism's height.
>
> - Decompose the prism into two or more smaller prisms of the same height and add the volumes of those smaller prisms.

Name _____   Date _____

A triangular prism has a rectangular prism cut out of it from one base to the opposite base, as shown in the figure. Determine the volume of the figure, provided all dimensions are in millimeters.

Is there any other way to determine the volume of the figure? If so, please explain.

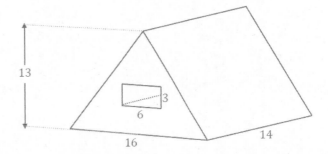

1. Find the volume of the three-dimensional object composed of right rectangular prisms.

   Volume of top and bottom prisms:

   $2(14 \text{ in.} \times 14 \text{ in.} \times 4 \text{ in.}) = 1,568 \text{ in}^3$

   Volume of middle prism:

   $5 \text{ in.} \times 5 \text{ in.} \times 10 \text{ in.} = 250 \text{ in}^3$

   Total volume:

   $1,568 \text{ in}^3 + 250 \text{ in}^3 = 1,818 \text{ in}^3$

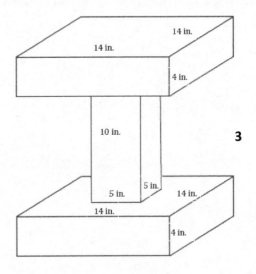

   > Similar to surface area, I can find the volume of each part of the object and then calculate the total volume by finding the sum of all the partial volumes.

2. Two students are finding the volume of a prism with a rhombus base but are provided different information regarding the prism. One student receives Figure 1 while the other receives Figure 2.

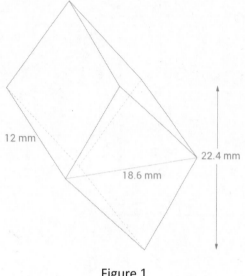

Figure 1

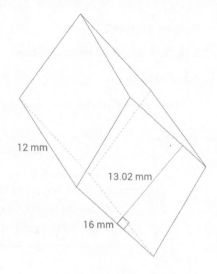

Figure 2

Find the volume in each case; show that the volumes are equal.

> In Figure 1, I split the base into two triangles in order to find the area of the base, and then I multiply by the height of the prism.

Volume of Figure 1: $2\left(\dfrac{1}{2}(18.6\text{ mm} \times 11.2\text{ mm})\right) \times 12\text{ mm} = 2{,}499.84\text{ mm}^3$

Volume of Figure 2: $16\text{ mm} \times 13.02\text{ mm} \times 12\text{ mm} = 2{,}499.84\text{ mm}^3$

> In Figure 2, I can use the information provided to find the area of the base and then multiply by the height of the prism.

EUREKA MATH

3. A plastic die cube for a game has an edge length of 2.5 cm. Throughout the cube, there are 15 cubic cutouts, each with an edge length of 3 mm. What is the volume of the cube?

> I know all sides of a cube are equal length.

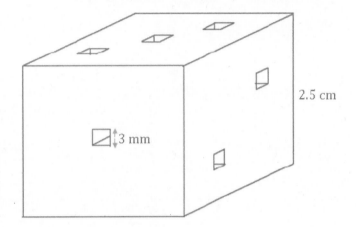

2.5 cm

3 mm

Volume of large cube:

$$(2.5 \text{ cm})^3 = 15.625 \text{ cm}^3$$

Volume of cutout cubes:

$$15(3 \text{ mm})^3 = 405 \text{ mm}^3$$

> I notice that the units for the two different volumes don't match. I convert mm^3 to cm^3 by dividing by 1,000.

Total volume of the die: $15.625 \text{ cm}^3 - 0.405 \text{ cm}^3 = 15.22 \text{ cm}^3$

> I find the total volume by subtracting the volume of the cutouts from the volume of the large cube.

4. A right rectangular prism has each of its dimensions (length, width, and height) increased by 20%. By what percent is its volume increased?

> When I increase each dimension by 20%, the new dimensions will be 100% + 20%, or 120%, of the original dimension.

$$V' = 1.2l \cdot 1.2w \cdot 1.2h$$
$$V' = 1.728lwh$$

The larger volume is 172.8% of the smaller volume, which means the volume increased by 72.8%.

> The smaller volume represents 100%, so the increase is the difference between the two percentages.

1. Find the volume of the three-dimensional object composed of right rectangular prisms.

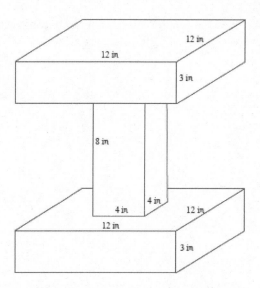

2. A smaller cube is stacked on top of a larger cube. An edge of the smaller cube measures $\frac{1}{2}$ cm in length, while the larger cube has an edge length three times as long. What is the total volume of the object?

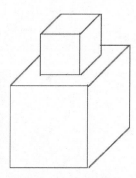

3. Two students are finding the volume of a prism with a rhombus base but are provided different information regarding the prism. One student receives Figure 1, while the other receives Figure 2.

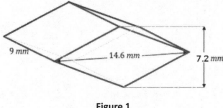

Figure 1

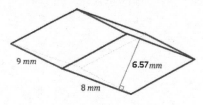

Figure 2

a. Find the expression that represents the volume in each case; show that the volumes are equal.

b. How does each calculation differ in the context of how the prism is viewed?

4. Find the volume of wood needed to construct the following side table composed of right rectangular prisms.

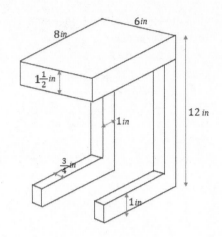

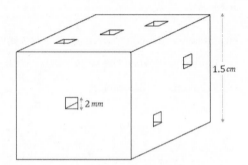

5. A plastic die (singular for dice) for a game has an edge length of 1.5 cm. Each face of the cube has the number of cubic cutouts as its marker is supposed to indicate (i.e., the face marked 3 has 3 cutouts). What is the volume of the die?

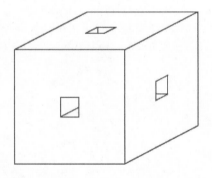

6. A wooden cube with an edge length of 6 inches has square holes (holes in the shape of right rectangular prisms) cut through the centers of each of the three sides as shown in the figure. Find the volume of the resulting solid if the square for the holes has an edge length of 1 inch.

7. A right rectangular prism has each of its dimensions (length, width, and height) increased by 50%. By what percent is its volume increased?

8. A solid is created by putting together right rectangular prisms. If each of the side lengths is increase by 40%, by what percent is the volume increased?

EUREKA MATH

Example 1

A swimming pool holds 10,000 ft³ of water when filled. Jon and Anne want to fill the pool with a garden hose. The garden hose can fill a five-gallon bucket in 30 seconds. If each cubic foot is about 7.5 gallons, find the flow rate of the garden hose in gallons per minute and in cubic feet per minute. About how long will it take to fill the pool with a garden hose? If the hose is turned on Monday morning at 8:00 a.m., approximately when will the pool be filled?

Example 2

A square pipe (a rectangular prism-shaped pipe) with inside dimensions of 2 in. × 2 in. has water flowing through it at a flow speed of $3\frac{\text{ft}}{\text{s}}$. The water flows into a pool in the shape of a right triangular prism, with a base in the shape of a right isosceles triangle and with legs that are each 5 feet in length. How long will it take for the water to reach a depth of 4 feet?

Exercise 1

A park fountain is about to be turned on in the spring after having been off all winter long. The fountain flows out of the top level and into the bottom level until both are full, at which point the water is just recycled from top to bottom through an internal pipe. The outer wall of the top level, a right square prism, is five feet in length; the thickness of the stone between outer and inner wall is 1 ft.; and the depth is 1 ft. The bottom level, also a right square prism, has an outer wall that is 11 ft. long with a 2 ft. thickness between the outer and inner wall and a depth of 2 ft. Water flows through a 3 in. × 3 in. square pipe into the top level of the fountain at a flow speed of $4\frac{\text{ft}}{\text{s}}$. Approximately how long will it take for both levels of the fountain to fill completely?

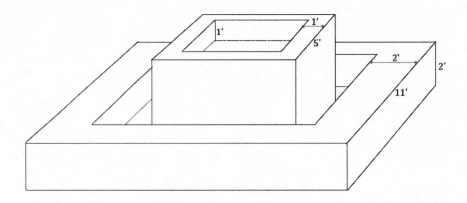

EUREKA MATH

Exercise 2

A decorative bathroom faucet has a 3 in. × 3 in. square pipe that flows into a basin in the shape of an isosceles trapezoid prism like the one shown in the diagram. If it takes one minute and twenty seconds to fill the basin completely, what is the approximate speed of water flowing from the faucet in feet per second?

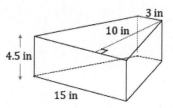

Lesson Summary

The formulas $V = Bh$ and $V = rt$, where r is flow rate, can be used to solve real-world volume problems involving flow speed and flow rate. For example, water flowing through a square pipe can be visualized as a right rectangular prism. If water is flowing through a 2 in. × 2 in. square pipe at a flow speed of $4\,\frac{\text{ft}}{\text{s}}$, then for every second the water flows through the pipe, the water travels a distance of 4 ft. The volume of water traveling each second can be thought of as a prism with a 2 in. × 2 in. base and a height of 4 ft. The volume of this prism is:

$$V = Bh$$

$$= \frac{1}{6}\,\text{ft.} \times \frac{1}{6}\,\text{ft.} \times 4\,\text{ft.}$$

$$= \frac{1}{9}\,\text{ft}^3$$

Therefore, $\frac{1}{9}\,\text{ft}^3$ of water flows every second, and the flow rate is $\frac{1}{9}\,\frac{\text{ft}^3}{\text{s}}$.

EUREKA MATH

Name _____ Date _____

Jim wants to know how much his family spends on water for showers. Water costs $1.50 for 1,000 gallons. His family averages 4 showers per day. The average length of a shower is 10 minutes. He places a bucket in his shower and turns on the water. After one minute, the bucket has 2.5 gallons of water. About how much money does his family spend on water for showers in a 30-day month?

1. Olivia has a leak in her new roof, so she puts a container in the shape of a right rectangular prism under the leak. Rainwater is dripping into the container at an average rate of 14 drops per minute. The container Olivia places under the leak has dimensions of 6 cm × 4 cm × 9 cm. Assuming each rain drop is roughly 1 cm^3, approximately how long does Olivia have before the container overflows?

   Volume of the container:

   $$6 \text{ cm} \times 4 \text{ cm} \times 9 \text{ cm} = 216 \text{ cm}^3$$

   Number of minutes until the container is filled with rainwater:

   $$216 \text{ cm}^3 \left(\frac{1 \text{ min.}}{14 \text{ cm}^3} \right) \approx 15.43 \text{ min.}$$

   > I determine the volume of the container and then use the rate to determine how long Olivia has until the container overflows.

   The bucket will overflow in about 15.43 minutes.

2. A basement flooded and contains $10,000 \text{ ft}^3$ of water that needs to be drained. At 1:00 p.m., a pump is turned on that drains water at the rate of 9 ft^3 per minute. Four hours later, at 5:00 p.m., a second pump is activated that drains water at the rate of 5 ft^3 per minute. At what time will the basement be free of water?

   Water drained during the first four hours:

   $$\left(\frac{9 \text{ ft}^3}{1 \text{ min.}} \right) (240 \text{ min.}) = 2,160 \text{ ft}^3$$

   > Once I determine how much water is drained when only one pump is working, I can determine how much water is left in the basement.

   Volume of water that still needs to be drained:

   $$10,000 \text{ ft}^3 - 2,160 \text{ ft}^3 = 7,840 \text{ ft}^3$$

   Amount of time needed to drain the remaining water:

   > When both pumps are working, I know the rate is 14 ft^3 per minute because I add the two rates together.

   $$\left(7,840 \text{ ft}^3 \right) \left(\frac{1 \text{ min.}}{14 \text{ ft}^3} \right) = 560 \text{ min.}$$

   It will take $13\frac{1}{3}$ hours to drain the basement, which means the basement will be free of water at 2:20 a.m.

3. A pool contains 12,000 ft³ of water. Pump A can drain the pool in 10 hours, and Pump B can drain the pool in 15 hours. How long will it take both pumps working together to drain the pool?

Rate at which Pump A drains the pool: $\frac{1}{10}$ *pool per hour*

Rate at which Pump B drains the pool: $\frac{1}{15}$ *pool per hour*

> I can determine the rate at which each pump drains the pool in order to determine the rate the water drains when both pumps are working together.

> I remember to find common denominators before adding the fractions.

Together, the pumps drain the pool at $\left(\frac{1}{10} + \frac{1}{15} \right)$ *pool per hour, or* $\frac{1}{6}$ *pool per hour. Therefore, it will take 6 hours to drain the pool when both pumps are working together.*

4. A 1,500-gallon aquarium can be filled with water flowing at a constant rate in 6 hours. When a decorative rock is placed in the aquarium, it can be filled in 5.25 hours. Find the volume of the rock in cubic feet (1 ft³ = 7.48 gal.).

Rate of the water flow into aquarium:

$$\frac{1{,}500\,\text{gal.}}{6\,\text{hours}} = \frac{250\,\text{gal.}}{1\,\text{hour}}$$

Volume of the rock in gallons:

$$\left(\frac{250\,\text{gal.}}{1\,\text{hour}} \right)(0.75\,\text{hour}) = 187.5\,\text{gal.}$$

> When the rock is placed in the aquarium, I know that it takes 0.75 hours less to fill. I can use the unit rate and time to determine the volume of the rock.

Volume of the rock in cubic feet:

$$(187.5\,\text{gal.})\left(\frac{1\,\text{ft}^3}{7.48\,\text{gal.}} \right) \approx 25.07\,\text{ft}^3$$

The volume of the rock is approximately **25.07 ft³**.

> In order to answer the question, I need to convert the volume of the rock from gallons to cubic feet.

Lesson 27:   Real-World Volume Problems

EUREKA MATH®

1. Harvey puts a container in the shape of a right rectangular prism under a spot in the roof that is leaking. Rainwater is dripping into the container at an average rate of 12 drops a minute. The container Harvey places under the leak has a length and width of 5 cm and a height of 10 cm. Assuming each raindrop is roughly 1 cm^3, approximately how long does Harvey have before the container overflows?

2. A large square pipe has inside dimensions 3 in. × 3 in., and a small square pipe has inside dimensions 1 in. × 1 in. Water travels through each of the pipes at the same constant flow speed. If the large pipe can fill a pool in 2 hours, how long will it take the small pipe to fill the same pool?

3. A pool contains 12,000 ft^3 of water and needs to be drained. At 8:00 a.m., a pump is turned on that drains water at a flow rate of 10 ft^3 per minute. Two hours later, at 10:00 a.m., a second pump is activated that drains water at a flow rate of 8 ft^3 per minute. At what time will the pool be empty?

4. In the previous problem, if water starts flowing into the pool at noon at a flow rate of 3 ft^3 per minute, how much longer will it take to drain the pool?

5. A pool contains 6,000 ft^3 of water. Pump A can drain the pool in 15 hours, Pump B can drain it in 12 hours, and Pump C can drain it in 10 hours. How long will it take all three pumps working together to drain the pool?

6. A 2,000-gallón fish aquarium can be filled by water flowing at a constant rate in 10 hours. When a decorative rock is placed in the aquarium, it can be filled in 9.5 hours. Find the volume of the rock in cubic feet (1 ft^3 = 7.5 gal.)

Credits

Great Minds® has made every effort to obtain permission for the reprinting of all copyrighted material. If any owner of copyrighted material is not acknowledged herein, please contact Great Minds for proper acknowledgment in all future editions and reprints of this module.